Illustrator:
Bruce Hedges

Editor:
Stephanie Buehler, M.P.W., M.A.

Editorial Project Manager:
Karen J. Goldfluss, M.S. Ed.

Editor in Chief:
Sharon Coan, M.S. Ed.

Art Director:
Elayne Roberts

Associate Designer:
Denise Bauer

Cover Artist:
Marc Kaslauskas

Product Manager:
Phil Garcia

Imaging:
Ralph Olmedo, Jr.

Publishers:
Rachelle Cracchiolo, M.S. Ed.
Mary Dupuy Smith, M.S. Ed.

Instant Science

primary

Author:
Julia Jasmine, M.A.

Teacher Created Materials, Inc.
P.O. Box 1040
Huntington Beach, CA 92647
ISBN-1-57690-063-0

Table of Contents

Introduction

Use *Instant Science* to perk up your science program and give your students something fun to do and think about every single day. This book contains age-appropriate natural science lessons for 50 weeks of the year. Each lesson includes five activities—a full week of instant science! As a bonus each lesson contains a language arts component to supplement your regular curriculum.

Instant Science can be used in two ways: by week (work straight through the book) or by subject (consult the index).

Instant Science accommodates several needs. It provides . . .

. . . fresh material for experienced teachers.

. . . a ready resource for beginning and substitute teachers.

. . . 50 weeks of science material for use in the increasingly popular year-round school—a need not addressed by traditional curriculum resources.

How to Use This Book

Building a Knowledge Base of Correct Information

Many of the activities in this book are designed to help students build a knowledge base by introducing them to scientific facts that will support their learning in the years to come. Recent research conducted by Howard Gardner *(The Unschooled Mind: How Children Think and How Schools Should Teach,* Basic Books, 1991) shows that children develop many misconceptions about the world around them simply because they do not have enough information. Although mastery of these concepts cannot be expected at the primary level, exposure to them will lay the groundwork for more effective scientific thinking later on.

Setting Up Portfolios

Portfolios are made to order for primary science. They give you an opportunity to view a student's collected work so that you can assess academic growth over a period of time. Simply staple together a week's work and allow the student to add a dated "Reflection" on the top page. There is no need to fuss about filing—just put each student's work into individual folders or have students file the work themselves. When you want to take a look at the work, you can quickly sort a folder full of stapled packets into chronological order and evaluate a student's progress.

Planning for Reflections

When students reflect on their work, they take control of the learning process because they observe and appreciate their own progress. Weekly reflections help students become at ease with the process of learning. It is also rewarding to take time for an overall reflection every month or six weeks. Students are usually astounded by their own growth.

Accommodating Different Intelligences

If you have a dominant *linguistic* intelligence, you may find that you teach—and expect students to learn—by reading, writing, and listening. It is helpful to remember that people learn in different ways:

- People with *spatial intelligence* need to see things. (Show lots of pictures and videos.)
- People with *logical-mathematical* intelligence like to collect and analyze data. (Point out the relationships among things.)
- People with *bodily-kinesthetic* intelligence need to touch things and do things. (Have lots of hands-on activities.)
- People with *musical intelligence* need music and rhythm. (Play music in the background and let them make up songs and rhythms to help themselves learn.)
- People with *interpersonal intelligence* need to interact with other people. (Use cooperative learning groups and let them discuss their ideas with others.)
- People with *intrapersonal* intelligence need time to think. (Let them do at least some things by themselves.)

Background

Help your students orient themselves in the universe with an expanded view of where they live, what their home is made of, and how it looks from space.

Day 1—What's Your Address?

Materials

- "What's Your Address?" form for each student (See below.)

Activity

Help students fill out the form. Discuss each entry, using maps and globes as needed.

Name ______________________________

Street ______________________________

City or Town ______________________________

County ______________________________

State ______________________________

Country ______________________________

Continent ______________________________

Hemisphere ______________________________

Planet ______________________________

System ______________________________

Galaxy ______________________________

Day 2—The Earth Is Like an Orange

Materials

- several thick-skinned oranges (Save them for Day 4.)
- a globe

Activity

Pass some thick-skinned oranges around the class. Have students run their fingers over the skin (Earth's "crust") and note the blossom and stem ends ("poles"). Ask them to close their eyes and imagine Earth as a round ball in space.

Show them a globe, pointing out the United States and your own state. Leave the globe in an accessible place for students to study.

Day 3—Inside the Earth

Materials

- two or three peaches—not too ripe
- world globe

Activity

Cut one or more peaches in half, leaving in the pit.

Say to students:

The peach has a thin skin. It would be more like the Earth if the skin were thick like an orange. There is material called the pulp (the part of the peach we eat) beneath the skin and a pit at its center. Earth has a center, too. It is called the "core."

Earth's core is very hot. It is made of melted rock. It heats the rock in the layers above it until some of it melts, too. Once in awhile the melted rock comes bursting out. We call the melted rock "lava" and the hole it comes out a "volcano."

Encourage questions and discussion.

Day 4—The Earth Is Wrapped in Air

Materials

- several thick-skinned oranges (from Day 2)
- a roll of cotton batting

Activity

Wrap each orange in a blanket of cotton. (You can use cotton balls, but they don't work as well.) Let students handle the wrapped oranges. Explain that air (atmosphere) is wrapped around the Earth and travels with the Earth through space. When we look up at the sky, we are looking up through Earth's blanket of air, which appears to us as the color blue. When astronauts in space look back at the Earth or take pictures of it, Earth looks blue. That is because they are looking down at the blanket of air (our sky) from the outside.

Day 5—Review and Reflect

Materials

- a video showing pictures of Earth taken from space
- "Review and Reflect" form (See below.)

Activity

Show the video. Review and discuss concepts introduced during the week. Help students complete the following form.

Review and Reflect

Name ______________________________ Date ______________

What I Learned:

I live on the outside surface of the ______________________________.

It has a thick skin called the ______________________________.

Inside, it has a middle called a ______________________________.

The middle of it is very ______________________________.

The outside is wrapped in a blanket of ______________________________.

The most interesting thing I learned was ______________________________

____.

Background

Help your students visualize the movement of the Earth that causes day and night.

Day 1—A Toy Top

Materials

- several small manual toy tops

Activity

Let all students take a turn spinning the tops. Point out that tops spin around a center that runs from tip to tip. Ask students to imagine a line going from top to bottom right through the middle. Tell students that this imaginary line is called the "axis." The turning movement around the axis is called its "rotation." Write these two words on the board and use them in context whenever appropriate.

Tell your students to pretend the top is the Earth. If the top were the Earth, the spinner end of the axis would be the North Pole and the bottom end, the end the top spins on, would be the South Pole.

Day 2—Your Turn to Turn

Materials

- a table lamp that casts light in all directions (not a directional desk lamp)

Activity

Ask for a student volunteer to play the part of Earth.

Tell students:

> The lamp is the sun. ________________is the "Earth."
>
> Think of the tops we used yesterday. What should the "Earth" (indicating student) be doing? (turning/rotating)

Have the "Earth" turn, stopping during discussions to prevent dizziness.

> Where is the "Earth's" North Pole? (top of head)
>
> Where is the "Earth's" South Pole? (feet)
>
> Where is the "Earth's" axis? (an imaginary line through body from head to feet)

Let as many students as possible volunteer to be "Earth."

Day 3—What About the Moon?

Materials

- a table lamp that casts light in all directions (not a directional desk lamp)
- a playground ball, soccer ball, or basketball

Activity

Ask for a student volunteer to play the part of "Earth."

Say to students:

The lamp is the sun. ________________ is the "Earth." But what about the moon?

This ball is the moon. (Have a student volunteer hold the moon.)

The moon goes around the Earth. (Have the "moon" walk slowly around the "Earth" as it begins to turn or rotate.)

The moon goes much more slowly than the Earth. It goes around the Earth only once for every 28 turns (or rotations) the Earth makes.

Let as many students as possible play the parts of the Earth and the Moon.

Day 4—Day and Night

Materials

- a table lamp that casts light in all directions (not a directional desk lamp)
- a playground ball, soccer ball, or basketball

Activity

Choose student volunteers to be the "Earth" and the "Moon" and have them start turning and walking around. Darken the room.

Say to students:

The lamp is the sun, __________ is the "Earth," and ____________ is the "Moon."

Pretend we live on the side of the Earth that _____________'s face is on.

Stop the movement when Earth turns his or her back on the lamp.

What is happening now? (It is night where we live.)

Stop the movement when Earth faces the lamp.

What is happening now? (It is day where we live.)

Let as many students as possible play the parts of the Earth and the Moon.

Day 5—Review and Reflect

Materials

- a table lamp that casts light in all directions (not a directional desk lamp)
- a globe
- "Review and Reflect" form (See below.)

Activity

Use the globe and the lamp to review and discuss the concepts introduced during the week. Help students complete the form.

Review and Reflect

Name ____ Date ________________

What I Learned:

The imaginary line running through the Earth is called its ________________________ .

It runs through the Earth from the ____ to

the ____.

The movement of the Earth turning around this imaginary line is called

____.

This movement of the Earth gives us________________and ________________________.

The ____ goes around the Earth.

While the Earth turns 28 times, the moon goes around only ________________________.

If the Earth did not turn ____

____

____

____

____

____.

Background

Help your students picture Earth as part of the solar system, its "family" of planets.

Day 1—Earth's Orbit; Mercury and Venus

Materials

- big cardboard circles labeled as sun, Mercury, Venus, Earth

Activity

Ask a student volunteer to be the "sun" and to stand in the middle of the room. Ask students to be the Earth and the moon, too. Have the Earth and moon stand as far away from the sun as possible. Remind the students about how the Earth and moon move and have them begin. Then, as the Earth and moon move together, ask them to also move around the perimeter of the room, circling the sun. All these movements are hard to do, so don't expect students to keep it up for long.

Say to students:

> While the Earth rotates on its axis and the moon goes around the Earth, they are both in orbit around the sun. A planet's orbit is its pathway around the sun. Each of these orbits is called a *revolution*. The Earth revolves around the sun. One complete revolution, or trip in orbit around the sun, takes about 365 days, or a year.

Give "Venus" and "Mercury" circles to two more student volunteers and explain:

> While the Earth and its moon are circling the sun, Mercury and Venus circle it, too. (Mercury is closest to the sun, and Venus is between Mercury and Earth.)

Ask students to stop their movement while you talk a little bit about Mercury and Venus and draw the diagram below on the chalkboard.

> Mercury is the closest planet to the sun. It was long thought to be the hottest planet, but scientists now think Venus is hotter.
>
> Venus is the planet between Mercury and Earth. Its thick clouds, made of material that would be poisonous to the people of Earth, hold the heat in.

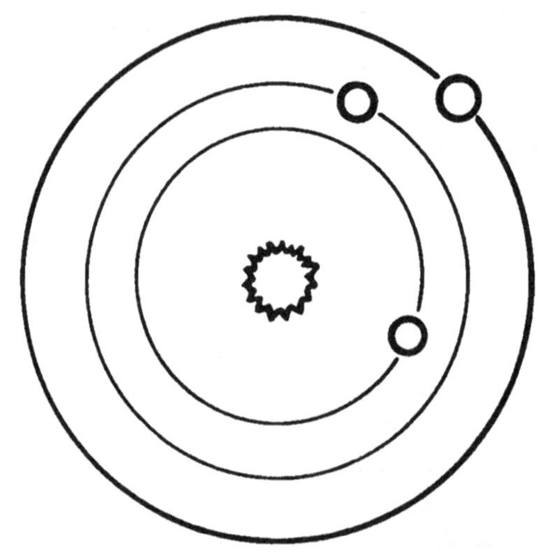

Let as many students as possible play the parts of the sun, the Earth and the moon, Mercury, and Venus.

Day 2—More Planets

Materials

- big cardboard circles labeled as sun, Mercury, Venus, Earth, Mars
- hula hoops for students to hold as "rings" of Saturn and Uranus

Activity

Expand yesterday's diagram on the board to include all the planets. (See the illustration.)

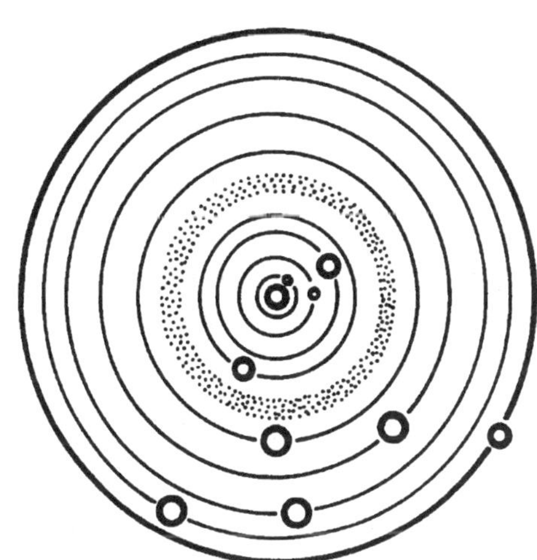

Move your solar system outside and add Mars, Jupiter, Saturn, Uranus, Neptune, and Pluto.

Have all the planets orbit the sun.

Let as many students as possible play the parts of the sun and the planets.

Day 3—The Asteroid Belt

Materials

- big cardboard circles labeled as sun, Mercury, Venus, Earth, Mars, Jupiter, Saturn, Uranus, Neptune, and Pluto
- hula hoops for students to hold as "rings" of Saturn and Uranus
- enough smaller cardboard circles labeled as "asteroid" so everyone in the class can participate at once

Activity

Expand yesterday's diagram on the board to include the asteroids. (See the illustration.)

Move your solar system outside and add the asteroids between Mars and Jupiter.

Have all the planets and the asteroids orbit the sun.

Let as many students as possible play the parts of the sun, all the planets, and the asteroids.

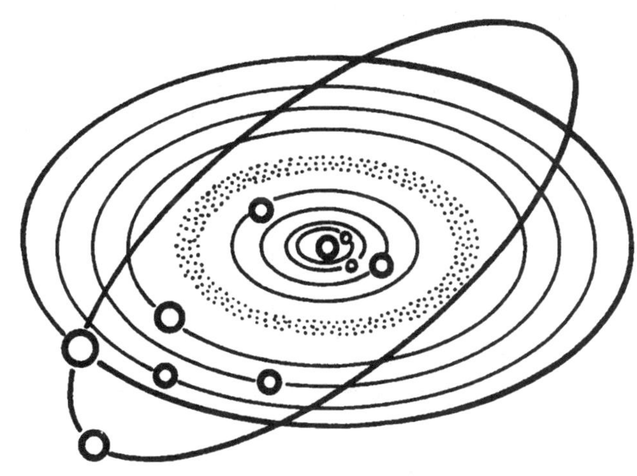

Day 4—Remembering the Names

Materials

- names of the planets written on the board or listed on a chart

Activity

Practice the names of the planets in order from the sun.

Tell students about ways of remembering things with clever sayings (mnemonic devices).

Write the mnemonic below on the board or overhead projector.

Say to students:

One way to remember the order of the planets from the sun is with a saying like this: **M**y **V**ery **E**nergetic **M**other **J**ust **S**erved **U**s **N**ine **P**izzas. Point out that the first letter of each word in the sentence is the same as the letter of a corresponding planet. Allow students who wish to make up their own.

Day 5—Review and Reflect

Materials

- "Review and Reflect" form (below)

Activity

Review and discuss concepts introduced during the week. Help students complete the form.

Review and Reflect

Name ______________________________ Date ______________

What I Learned:

The Earth and all the other planets revolve around the ______________________.

Their paths are called ______________________________________.

The names of the planets are ________________, ________________,

________________, ________________, ________________,

________________, ________________, ________________,

and ________________.

Background

Give your students a quick introduction to the other planets in the solar system.

Day 1—The Inner Planets

Materials

- current pictures and/or videos of the inner planets (Mercury, Venus, Earth, Mars)

Activity

Take plenty of time to look at pictures and/or videos. Talk about each of the inner planets in turn. Share some of the facts, below, to give your students some information about each planet. You may wish to reproduce this list for your students.

Mercury

- Mercury is the second hottest planet and the one closest to the sun.
- It takes 88 days to go around the sun.
- It rotates around its axis only once every 59 Earth days so its day is 59 Earth days long.
- Mercury is just a little bit bigger than Earth's moon.
- It has neither moons of its own nor an atmosphere.

Venus

- Venus is the second planet from the sun.
- Venus is very hot with a thick atmosphere of poisonous gases.
- It is about the same size as Earth.
- It takes 225 days to go around the sun.
- It rotates once every 243 Earth days.
- Venus has no moons.

Earth

- Earth is the third planet from the sun.
- Earth's atmosphere protects it from much of the sun's heat and harmful rays.
- Over 70 percent of the Earth is covered by water.
- Earth rotates once every 24 hours and takes about 365 days to go around the sun.
- Earth has one moon.

Mars

- Mars is the fourth planet from the sun.
- It has a thin atmosphere and some water trapped in its polar ice caps.
- Mars is about half the size of Earth.
- Its day (the time it takes to rotate about its axis) is about the same as Earth's.
- It takes 687 Earth days to go around the sun.
- Mars has two small moons.

Day 2—Jupiter and Saturn

Materials

- current pictures and/or videos of Jupiter and Saturn

Activity

Take plenty of time to look at pictures and/or videos. Share some of the facts, below, to give your students some information about each planet. You may wish to reproduce this list for your students.

Jupiter

- Jupiter is the fifth planet from the sun.
- It is the largest planet in our solar system and has at least 16 moons. It is almost like a solar system on a small scale.
- The planet is made up almost entirely of liquid hydrogen which gets very hot near the center.
- Jupiter rotates very fast. Its day lasts only 10 Earth hours.
- Jupiter's fast spinning combined with its atmosphere results in constant storms.
- Scientists think that Jupiter's red spot is a storm that has lasted almost 300 years.
- It takes Jupiter almost 12 Earth years to orbit the sun.

Saturn

- Saturn is the sixth planet from the sun.
- Although Jupiter, Uranus, and Neptune have rings too, Saturn's rings are the biggest and brightest. Scientists think they are really orbiting chunks of rock and ice.
- Saturn has around 20 moons. The largest, Titan, has an atmosphere of its own.
- Saturn's atmosphere is very much like Jupiter's.
- Saturn rotates in 10 Earth hours and takes 29½ Earth years to orbit the sun.

Day 3—The Outer Planets

Materials

- current pictures and/or videos of Uranus, Neptune, and Pluto

Activity

Take plenty of time to look at pictures and/or videos. Share some of the facts, below, to give your students some information about these planets. You may wish to reproduce this list for your students.

Day 3—The Outer Planets *(cont.)*

Uranus

- Uranus is the seventh planet from the sun.
- Uranus rotates once every 17 Earth hours and takes 84 Earth years to travel around the sun.
- It has a thin ring around its equator and 15 moons.

Neptune

- Neptune is usually the eighth planet from the sun, but sometimes its orbit crosses Pluto's orbit and it becomes the ninth planet for awhile.
- Neptune rotates every 18 Earth hours and takes 165 Earth years to travel around the sun.
- Its atmosphere is made of hydrogen and helium, and it has eight moons.

Pluto

- Pluto is usually the ninth planet from the sun. It is so far away that the sun would appear to be just a bright star from its surface.
- It has one moon of its own, Charon, which is about half its size.
- Pluto rotates every 6½ Earth hours and takes 248 Earth years to travel around the sun.

Day 4—The Asteroid Belt

Materials

- pictures and/or videos of asteroids

Activity

Take plenty of time to look at pictures and/or videos. Talk about the asteroids. Choose from these facts to give your students some information about them. You may wish to reproduce this list for your students.

Asteroids

- The asteroid belt consists of thousands of tiny planets in an orbit between Mars and Jupiter.
- The largest, Ceres, is only about 600 miles in diameter. Scientists are not sure how they originated. Some think they are material that didn't pull together to make a planet when the solar system formed.

Day 5—Review and Reflect

Materials

- paper and pencil

Activity

Review and discuss concepts introduced during the week. Then, direct students to write one interesting fact about each of the planets.

Background

Take this opportunity to review and reinforce the concepts and vocabulary introduced during the first four weeks.

Day 1—Where We Live

Materials

- an orange and a peach
- a globe
- cotton batting

Activity

Use the following talk to question students and prompt their review of concepts. Write important vocabulary words on the board as students answer your questions.

Say to students:

Here is an orange. It has a skin. If I hold it this way, the stem end is on the top and the blossom end is on the bottom. Pretend that this orange is the Earth. What is the Earth's "skin" called? (crust)

What is the top of the Earth called? (North Pole)

What is the bottom of the Earth called? (South Pole)

Now look at the peach. In the center of this peach is a pit. What is the Earth's center called? (core)

Now I'm going to wrap cotton around the orange. What is the Earth wrapped in? (atmosphere or air)

Take an additional moment to look at the globe and find the place where you live to review the concepts introduced in Week 1, Day 1.

Day 2—Turning Like a Top

Materials

- toy tops
- lamp

Activity

Use the following talk to question students and prompt their review of concepts. Write important vocabulary words on the board as students answer your questions.

Let's spin these tops and remember what we learned about the Earth and its moon.

What is the imaginary line running through the Earth from top to bottom called? (axis)

What is the Earth's movement when it turns around this imaginary line called? (rotation)

What does this movement of the Earth give us? (night and day)

What goes around the Earth? (moon)

Who wants to be first to show us how the Earth moves?

Day 3—Going Around the Sun

Materials

- big cardboard circles labeled as sun and planets: sun, Mercury, Venus, Earth, Mars, Jupiter, Saturn, Uranus, Neptune, and Pluto
- hula hoops for rings of Saturn and Uranus
- enough smaller cardboard circles lettered with the word "asteroid" so everyone in the class can participate at once

Activity

Use the following questions to prompt their review of concepts. Write important words on the board, including the names of the planets, as students answer your questions.

> Let's see if we can remember the names of all the planets in order from the sun. (Mercury, Venus, Earth, Mars, Jupiter, Saturn, Uranus, Neptune, Pluto)
>
> What did you do to help yourself remember them? Did you use a silly sentence? Did you make up your own silly sentence?
>
> All of the planets travel around the sun. What is the word that means the path they take? (orbit)
>
> Where does the asteroid belt fit in? (between Mars and Jupiter)
>
> Now we can take our circles with the names of the sun and all the planets and asteroids and go out on the playground and take turns orbiting the sun.

Day 4—Planet Closeups

Materials

- current video or pictures of the planets
- picture or diagram of the solar system

Activity

Show the video or pictures and discuss what was learned about the various planets.

Ask students to volunteer facts that they remember. Go through the planets one by one, asking questions such as the following:

> Can anyone tell me something about Mercury?
>
> What is the fact about Venus that you found most interesting?

Day 5—Review and Reflect

Materials

- drawing paper
- pens, pencils, crayons, markers
- "Review and Reflect" form (below)

Activity

Have each student draw a picture of the solar system, labeling the planets. Accept all versions, as they will give you a real look at each child's concept level.

Give students necessary help to complete the form below. Depending on the age group you are teaching, they can do it themselves or dictate their information to you or to an aide.

Review and Reflect

Name ______________________________ Date ______________

Choose words from the list below to complete this story.

I live on the ______________. It is a ______________

that travels in an ______________ around the ______________.

The Earth is one of ______________ planets in our ______________.

The Earth rotates around its own ______________.

This gives us ______________ and ______________.

night	day	Earth
nine	orbit	axis
planet	solar system	sun

Background

Acquaint your students with some facts about Earth's atmosphere.

Day 1—The Gas We Breathe

Materials

- ice
- hot plate and pan

Activity

Say to students:

You already know that the Earth is wrapped in the air that we breathe. This air is a mixture of gases. It is transparent. We can see right through it.

Gases are one kind of matter. The other two types of matter are solids and liquids.

Water comes in all three of the forms of matter. It is usually a liquid; we can pour it.

When it freezes, it becomes a solid; we call it ice. When it is heated, it becomes a gas; we call it steam or water vapor.

Pour some ice cubes into a pan and heat them. Let students see that the ice cubes have turned into liquid water. Keep heating the water until it boils and makes steam.

There is water vapor in the air we breathe. There are other gases, too. One of them is called oxygen. We need oxygen to live. When you breathe in (have everyone take a deep breath), you take in oxygen for your body to use. When you breathe out (have everyone exhale), you breathe out a gas called carbon dioxide.

Plants are different. They take in carbon dioxide and give off oxygen. We need each other to have the correct gases to live.

Day 2—Air Is Heavy

Materials

- paper or plastic cups for each student
- plastic drinking straws for each student
- water

Activity

Say to students:

The Earth's atmosphere (our blanket of air) is heavy. It presses down on us all the time. The air pressing down on you right now weighs about a ton. You don't feel it because you are also held up by the air that presses down against you.

When you use a straw to drink, you suck the air out of the straw, leaving a vacuum. The weight of the air pressing down on the water or other liquid pushes the water up to fill the vacuum. (Have everyone try this.)

Day 3—Warm Air Expands and Rises

Materials

- empty pop bottle
- hot water
- balloon
- ice
- pans

Activity

Say to students:

Warm air expands; that is, it fills more space than cool air.

This empty soda bottle is already full of air. I'm going to slip this balloon over the top of it. Watch what happens when I set it in a pan of hot water. (Balloon should inflate.)

Now watch what happens when I set it in a pan of ice water. (Balloon should deflate.)

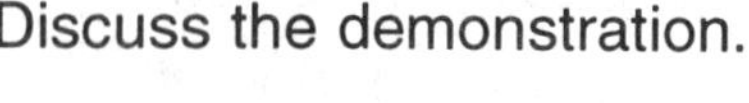

Discuss the demonstration.

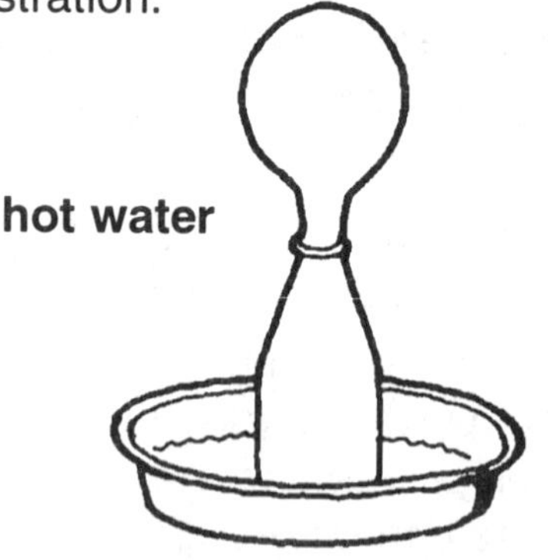

Day 4—Wind Is Moving Air

Materials

- construction paper, one piece for each student
- a globe

Activity

Have students pleat the construction paper to make fans.

Say to students:

We know this room is full of air. Wave your fans to make the air move. Moving air is called wind.

Demonstrate with a globe as you discuss the following:

We also know that warm air rises and expands. The Earth gets hottest around the equator. The warm air around the equator rises, and the colder air from the north and south moves in to take its place. This movement of air combined with the Earth's rotation causes wind to blow all over the Earth.

Discuss the demonstration.

Day 5—Review and Reflect

Materials

- a globe and other props from the week's demonstrations and experiments
- "Review and Reflect" form (below)

Activity

Use the globe and other props to review and discuss the concepts introduced during the week. Help students complete the form, dictating their answers to you or an aide, if needed. (When students have finished, see if anyone was able to make the inference about plants renewing our oxygen supply. Depending on the age group you are teaching, discuss the idea more fully.)

Review and Reflect

Name ______________________________ Date ______________

What I Learned:

Water is a liquid. It can also be a solid called ______________________.

Water can also be a ______________________ called water vapor.

Our air (Earth's atmosphere) is a mixture of gases.

We breathe in a gas called ______________________.

We breathe out a gas called ______________________.

Since people and animals have been breathing in oxygen and breathing out carbon dioxide for thousands and thousands of years, why do you think we haven't run out of oxygen?

____

____

____

____

____

____

____

Background

Acquaint your students with Earth's major landforms, the continents.

Day 1—One Big Landmass

Materials

- individual outline maps of the world

Activity

Tell students:

Many scientists think that all of the land on Earth once formed one huge supercontinent that split up and drifted apart.

They think this because of the shapes of the continents, which look as if they would fit together like a puzzle if they were moved close together.

Cut out the continents on your map and see what you think.

Let's compare our results and talk about them.

Day 2—Continental Closeups: North and South America

Materials

- classroom wall map of the world
- pictures and/or videos of landmark areas

Activity

Take plenty of time to look at pictures and/or videos. Talk about each of the continents in turn. Choose from this information to give your students some facts about each one.

North America

- North America looks a little bit like a huge triangle. The point at the bottom connects with South America.
- The Pacific Ocean, the Atlantic Ocean, and the Arctic Ocean are on the three sides.
- Great mountain ranges are separated by wide plains. The mountains run north and south near the coasts.
- The climate varies from very cold in the North to very hot in the South.

Day 2—Continental Closeups: North and South America *(cont.)*

South America

- South America looks a little bit like a huge triangle, too.
- The Pacific Ocean, the Atlantic Ocean, and the Caribbean Sea border it on three sides.
- The Andes Mountains run along the entire western coast.
- The equator runs across the broadest part of South America, and most of the continent is warm all year except in the Andes highland where it is always cold.

Day 3—Continental Closeups: Europe, Asia, and Africa

Materials

- classroom wall map of the world
- pictures and/or videos of landmark areas

Activity

Take plenty of time to look at pictures and/or videos. Talk about each of the continents in turn. From this information, choose some facts about each one to give your students.

Europe

- Europe is the second smallest continent and can be thought of as a peninsula joined to Asia at the Ural Mountains. Both continents together are sometimes called Eurasia.
- It is bordered on the north and west by the Arctic Ocean and the Atlantic Ocean and separated from Africa on the south by the Mediterranean Sea.
- There are many mountain ranges in Europe, including the Alps.
- The climate of Europe is mild because of the westerly winds that blow off the Atlantic and the warm ocean currents.

Asia

- Asia has a larger area in square miles and more people than any other continent.
- Asia is bordered on the north and east by the Arctic Ocean and the Pacific Ocean and on the south by the Indian Ocean. It is bordered on the west by Europe.
- Asia has the highest, lowest, coldest, hottest, wettest, and driest places on Earth.
- The climate of Asia varies, but much of it is controlled by the monsoon winds that bring hot, wet summers and cool, dry winters.

Africa

- Africa, the second largest continent, is bordered on the west by the Atlantic Ocean and on the east by the Indian Ocean. The Mediterranean Sea separates it from Europe on the north.
- Most of Africa is desert, forest, or grassland with only occasional mountains.
- Africa ranks with South America as the warmest of the continents. Some areas have very heavy rain and some areas, like the Sahara Desert, have almost none.

Day 4—Continental Closeups: Australia and Antarctica

Materials

- classroom wall map of the world
- a globe (for Antarctica)
- pictures and/or videos of landmark areas

Activity

Take plenty of time to look at pictures and/or videos. Talk about each of the continents in turn. Use the information below to give your students some facts about each one.

Australia

- Australia, the smallest continent, is the only continent that is also a country.
- It is bordered on the west by the Indian Ocean and the east by the Pacific Ocean.
- Australia is generally flat except for highlands along the eastern coast.
- Australia's climate is warm and dry.

Antarctica

- Antarctica is the continent surrounding the South Pole.
- It is bordered by what might be thought of as the "bottom" edges of the Atlantic, Pacific, and Indian Oceans.
- Most of Antarctica lies buried under mile-thick ice and snow. Mountains stick through in the interior and along the coasts.
- Antarctica is the coldest continent and the most desolate place on Earth. The only people living there are scientists who are studying the continent.

Day 5—Review and Reflect

Materials

- "Review and Reflect" form (below)
- paper and pencil

Activity

Review and discuss concepts introduced during the week. Help students complete the form.

Review and Reflect

Name ______________________________ Date ______________

There are ______________________________ continents.

On another piece of paper write one interesting fact about each continent.

Background

Acquaint your students with Earth's major bodies of water: oceans, lakes, and rivers.

Day 1—One Big Ocean

Materials

- a globe
- pictures and/or a video that shows the Earth from space, with good views of the oceans and continents

Activity

Take time to show the video or look at the pictures. Give everyone in the class a chance to take a good look at the globe to observe the relationship between water and land areas.

Say to students:

> About seven-tenths of the Earth is covered with water. That means if you divided the Earth's surface into ten parts, seven of them would be water and only three would be land.
>
> We talk about the different oceans by name—Atlantic, Pacific, Indian, and Arctic—but there is no way to truly tell where one stops and another starts.
>
> All the oceans run together into what could be thought of as one big ocean. Look at the globe and see for yourselves.
>
> Let's compare our results and talk about them.

Day 2—Ocean Closeups

Materials

- a globe
- classroom wall map of the world
- pictures and/or videos of landmark areas

Activity

Take plenty of time to look at pictures and/or videos. Talk about each of the oceans in turn. Use the information below to give your students some facts about each one.

Pacific Ocean

- The Pacific is the largest of the oceans. It is also the deepest.
- North and South America are on the east side of the Pacific Ocean; Asia and Australia are to its west. On the north, the Pacific is linked to the Arctic Ocean by the Bering Strait. On the south, the Pacific joins the Atlantic and Indian Oceans around Antarctica.
- The land around this ocean is known as the "Pacific Rim."

Day 2—Ocean Closeups *(cont.)*

Atlantic Ocean

- The Atlantic is the second largest of the oceans.
- North and South America are on the west side of the Atlantic Ocean and Europe and Africa on its east. It joins the Arctic Ocean on the north and the Pacific and Indian Oceans on the south.
- Until recently, it was the most important ocean for commerce.

Indian Ocean

- The Indian Ocean is not quite as large as the Atlantic Ocean.
- Africa borders the Indian Ocean on the west and Australia and the East Indies on the east. Asia is on the north. The Indian Ocean joins the Atlantic and Pacific on the south.
- The Indian Ocean is relatively calm although it is sometimes swept by typhoons.

Arctic Ocean

- The Arctic Ocean is the smallest of the oceans.
- It is located at the top of the world, north of North America, Europe, and Asia.
- It is covered with ice and snow during most of the year.

Day 3—Lovely Lakes

Materials

- a globe
- classroom wall map of the world
- pictures and/or videos of landmark areas
- "Looking Up Lakes" form, if desired (below)

Activity

Take plenty of time to look at pictures and/or videos. Talk about some of the lakes you see. Use the information below to give your students some facts about lakes or help students find information about them in reference books to complete the form below.

- A lake is a body of water surrounded by land.
- Some famous lakes are the Great Lakes in the United States, Lake Victoria in Africa, the Dead Sea in Asia, Loch Ness in Scotland, and Lake Titicaca in South America.
- The Dead Sea is the lowest lake in the world (1,286 feet/386 m below sea level), and Lake Titicaca is the highest (12,507 feet/3.8 km above sea level in the Andes).

Name ______________________________ Date ______________

Looking Up Lakes

Name of lake ______________________________

Country ______________________________

Interesting fact ______________________________

Day 4—Rising Rivers

Materials

- a globe
- classroom wall map of the world
- pictures and/or videos of landmark areas
- "Looking Up Rivers" form, if desired (below)

Activity

Take plenty of time to look at pictures and/or videos. Talk about some of the rivers you see. Use the information below to give your students some facts about rivers or help students find information about them in reference books to complete the form below.

- A river is a natural stream of water that flows into an ocean, a lake, or another river.
- Some famous rivers are the Missouri and the Mississippi in the United States, the Amazon in South America, the Danube in Europe, the Mekong in Asia, and the Nile in Africa.
- The Nile is the longest and the Amazon the second longest river in the world.

Name ______________________________ Date ______________

Looking Up Rivers

Name of river ______________________________

Country ______________________________

Interesting fact ______________________________

Day 5—Review and Reflect

Materials

- "Review and Reflect" form (below)
- paper and pencil

Activity

Review and discuss concepts introduced during the week. Help students complete the form, allowing them to dictate information to you or an aide if needed.

Review and Reflect

Name ______________________________ Date ______________

On another piece of paper, write one interesting fact about each of Earth's oceans.

Background

Acquaint your students with some facts about the interior of the Earth.

Day 1—Diagramming the Earth

Materials

- a globe
- a diagram of the Earth's interior for each student (See below.)
- crayons: red, orange, tan, and brown

Activity

Say to students:

This is what scientists think the inside of the Earth looks like. Starting at the center of the Earth, color the sections as follows:

Color the middle section red.

Color the second section orange.

Color the third section tan.

Color the outside section brown.

Save the colored diagram for Day 2.

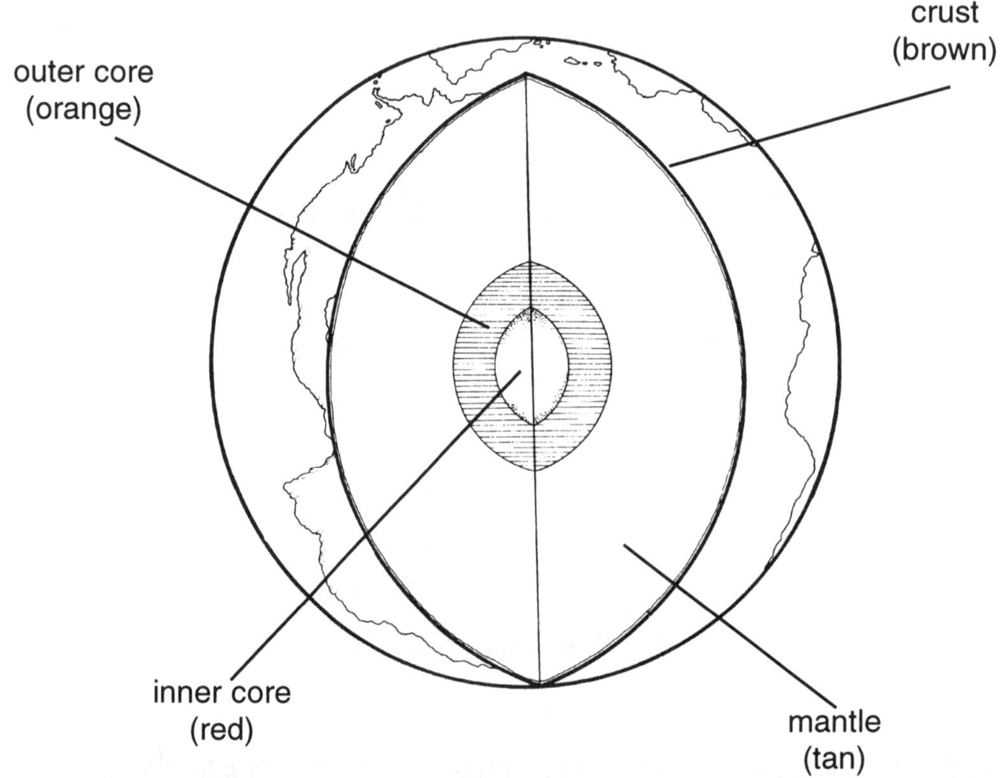

Day 2—Name the Layers

Materials

- a globe
- completed diagram of the Earth's interior for each student (from Day 1)
- pencils or pens

Activity

Say to students:

Scientists have special names for the inside layers of the Earth.

The part you colored red is the inner core.

The part you colored orange is the outer core.

The part you colored tan is the mantle.

The part you colored brown is the crust.

Write the underlined words on the board so students can label their diagrams. Have them write the words and then draw arrows to the layers.

Save the labeled diagram for Day 3.

Day 3—A Clay Model of the Earth

Materials

- a globe
- completed diagram of the Earth's interior for each student (from Day 1)
- play dough (purchased or homemade) in red, orange, tan, and brown

Activity

Say to students:

Today we are going to make a model of the Earth's interior. Use your labeled diagrams to help you.

Roll a small ball of red play dough for the inner core.

Cover the red ball with a thick layer of orange for the outer core and roll it until it is round and smooth.

Now cover the orange ball with a thick layer of tan play dough for the mantle.

Finally, add a thin layer of brown play dough to represent the crust.

Leave the Earth models on the students' desks so each person will have his or her own model to use for the activity on Day 4. (The diagram can go into the portfolios.)

Day 4—Cutaway View of the Earth

Materials

- a globe
- clay models of the Earth from Day 3
- dull knives such as plastic picnic knives or pumpkin-carving knives

Activity

Say to students:

It would be impossible to guess what the inside of the Earth looks like from the models that you have made, so we are going to cut a wedge out of the models to look at the inside. They should look like the diagram you colored and labeled.

Decide where the North Pole is on your model and cut through the clay all the way to the South Pole. (Cut a wedge. Do not cut the clay model in half.)

Then make another cut starting and ending in the same places so that you can lift out a wedge of clay.

How many of you found the inner core? The outer core? The mantle?

Take the point of your knife and drag it from the orange outer core up through the crust.

This shows what happens when the material from inside the Earth pushes its way through the mantle and comes out of a volcano.

Day 5—Review and Reflect

Materials

- "Review and Reflect" form (below)

Activity

Review and discuss concepts introduced during the week. Help students complete the form.

Review and Reflect

Name ______________________________ Date ______________

What I Learned:

The Earth is made up of ______________________________ layers.

The layers are called ______________________________

____

Background

Take this opportunity to review and reinforce the concepts and vocabulary introduced during the last four weeks.

Day 1—Interesting Air

Materials

- none

Activity

Ask students:

Who remembers the names of the three kinds of matter? (solid, liquid, gas)

What kind of matter is air? (gas)

Which gas do we breathe in? (oxygen)

Which gas do we breathe out? (carbon dioxide)

Which gas do plants take in? (carbon dioxide)

Which gas do plants give off? (oxygen)

Does air weigh anything? (yes)

What does warm air do? (expands and rises)

What do we call moving air? (wind)

Discuss each concept.

Day 2—Earth's Land

Materials

- a globe
- a classroom wall map of the world

Activity

Say to students:

Who can use the map or the globe to show me one of the continents? Point to it and then write its name on the board.

Who can tell us one fact about that continent?

Continue until you have covered all seven.

Day 3—Earth's Water

Materials

- a globe
- a classroom wall map of the world

Activity

Say to students:

Who can use the map or the globe to show me one of the oceans? Point to it and then write its name on the board.

Who can tell us one fact about that ocean?

Continue until you have covered all four.

Day 4—Inside the Earth

Materials

- clay models of the Earth from Week 9, Day 4

Activity

Say to students:

Take the wedge of clay out of your model of the Earth and look at the inside.

What is the name of the red part? (the inner core)

What is the name of the orange part? (the outer core)

What is the name of the tan part? (the mantle)

What is the name of the brown outside layer? (the crust)

From which layer does liquid rock or lava work its way to the surface? (the outer core)

What can you guess about the temperature of the Earth's insides? (hot)

What makes you think so? (Rock would have to get very hot before it would melt.)

Tell students they will learn more about melted rock and the temperature inside the Earth when they study volcanoes.

Day 5—Review and Reflect

Materials

- drawing paper
- pens, pencils, crayons, markers
- "Review and Reflect" form (below)

Activity

Have each student draw a picture (diagram) of the inside of the Earth, labeling the layers. Accept all versions; they will give you a good assessment of each child's conceptual level.

Give students necessary help to complete the form below.

Review and Reflect

Name ______________________________ Date ______________

Choose words from below to complete this story.

The Earth is wrapped in a blanket of ______________________________.

Air is a mixture of different ______________________________.

We breathe in the ______________________________ in the air.

We breathe out ______________________________ .

The surface of the Earth has both ______________ and ______________.

The land is divided into seven ______________________________.

Much of the water can be found in __________, __________ and __________ .

The inside of the Earth has four layers: ______________________________,

______________________________, ______________________________,

and ______________________________.

air	oxygen	lakes
crust	land	inner core
gases	oceans	carbon dioxide
mantle	water	continents
outer core	rivers	

Background

Use what your students have learned about the inside of the Earth to present some facts about earthquakes.

Day 1—Faults

Materials

- several blocks of wood, painted to show soil, earth, and rock layers and cut on a slant (See page 35.)

Activity

Say to students:

Now that we know something about the inside of the Earth, we can understand how earthquakes can happen.

Movement inside the Earth causes blocks of rock near the surface to break and then move along the crack.

This kind of crack is called a fault.

Rocks and soil may slip any direction along a fault—up, down, and sideways.

Demonstrate movement and then divide students into groups so they may experiment with the blocks.

Day 2—Effects of Earthquakes

Materials

- blue paint
- blocks from Day 1

Activity

Use blue paint to indicate a river on the top surfaces of the wooden blocks.

Say to students:

If an earthquake happens on a fault line that a river crosses, several things could happen.

Sideways movement could change the course of the river.

Up or down movement could create a waterfall or a lake.

Demonstrate movement and then divide students into groups so they may experiment with the blocks.

Day 2—Effects of Earthquakes *(cont.)*

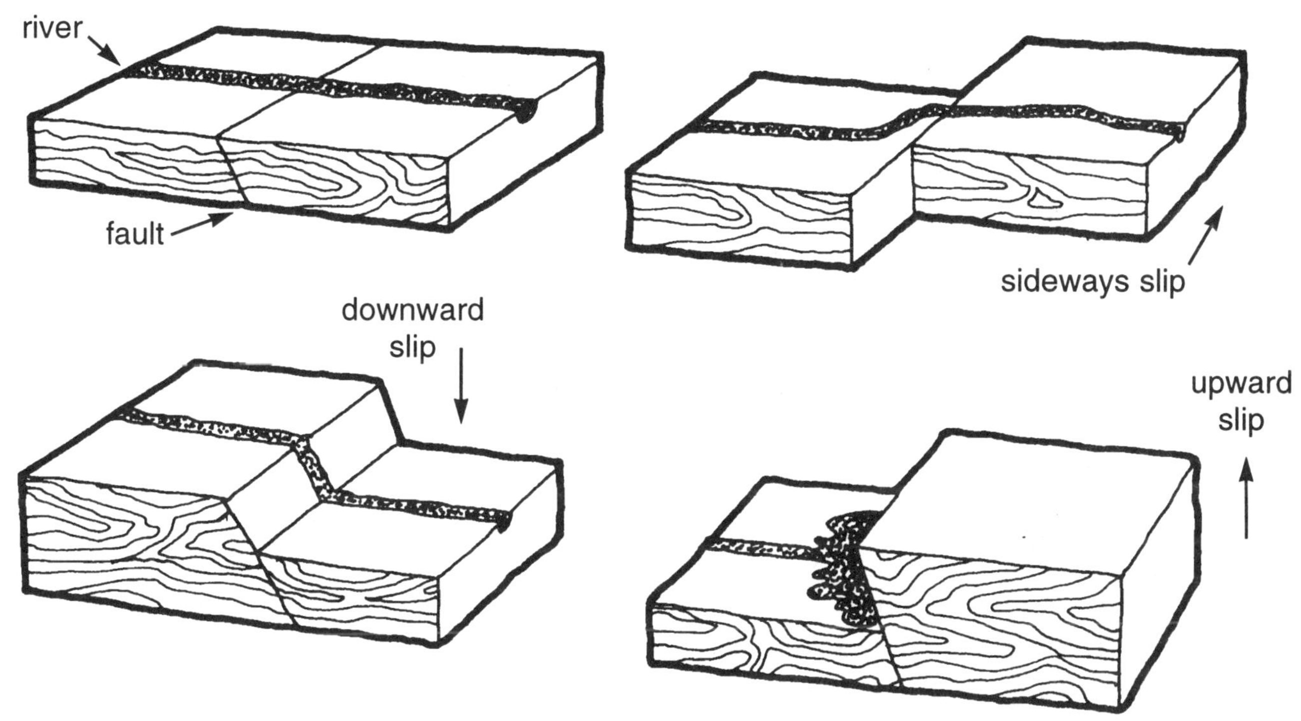

Day 3—Safety Drills

Materials

- none

Activity

Say to students:

Students in schools everywhere have fire drills. What do we do when we have a fire drill?

(Let student volunteers review your fire drill rules.)

In places where earthquakes happen, people have "earthquake drills." (If you have earthquake drills, let student volunteers review the rules.)

In an earthquake drill, the first step is to "duck and cover." This means to get under something like a desk or a table and cover the back of your neck with your hands.

The next step is to move out of the building after the shaking has stopped if there has been a real earthquake or after a signal if there has been a drill.

What other types of drills do we have? Who can tell us the rules we follow?

If time allows, you may want to have a practice drill of one kind of another.

Day 4—Getting Prepared

Materials

- flashlight
- radio with fresh batteries
- water purification tablets
- hat or cap with a brim
- backpack disaster kit containing items such as the following: bottle of water, nonperishable snacks, small game or two

Activity

If you already have a disaster program, review it with students. If not, *say to students*:

In some places, all students have an emergency kit like this one, ready to pick up from a closet near the door.

Why would you need a flashlight? (in case the lights go out)

Why would you need a radio? (to listen for information and instructions)

Why would you need water or water purification tablets? (in case the regular water supply is not safe to drink)

Why would you need snacks? (in case you get hungry)

Why would you need games? (to help pass the time while you wait)

Why should everything be packed in a backpack? (so you can carry it)

Preparing kits could be a good project for your class or for your school's parent organization.

Day 5—Review and Reflect

Materials

- "Review and Reflect" form (below)

Activity

Review and discuss concepts introduced during the week. Help students complete the form, allowing them to dictate to you or an aide if needed.

Review and Reflect

Name ______________________________ Date ______________

What I Learned:

Cracks in the Earth's surface caused by movement of rocks inside the Earth are called

______________. The ground around one of these cracks can move three ways:

______________, ______________, and ______________.

Background

Use what your students have learned about the inside of the Earth to present some facts about volcanoes.

Day 1—Review Inside the Earth

Materials

- diagrams of the inside of the Earth (from Week 9)
- clay models of the inside of the Earth (from Week 9)

Activity

Say to students:

Remember what we learned about the inside of the Earth. Look at your model and your diagram.

The rock in the outer core is so hot that it is melted. Another name for this melted rock is magma.

Sometimes the magma squeezes up through the rocks in the mantle and comes out through weak places in the Earth's crust.

Sometimes it just oozes out and runs down the volcano. Sometimes its path is blocked, and it builds up great pressure and finally explodes in all directions.

The explosion is called an eruption.

When the magma cools, it is called lava.

Day 2—Cutaway Diagram of a Volcano

Materials

- cutaway diagram of a volcano for each student (enlarge and copy diagram on page 38)
- large labeled diagram for display (see page 38)

Activity

Say to students:

This is a sample diagram of the inside of a volcano.

Label your diagram to match the large one.

A conduit is another name for something shaped like a pipe.

The crater is the hole where the magma comes out.

Day 2—Cutaway Diagram of a Volcano *(cont.)*

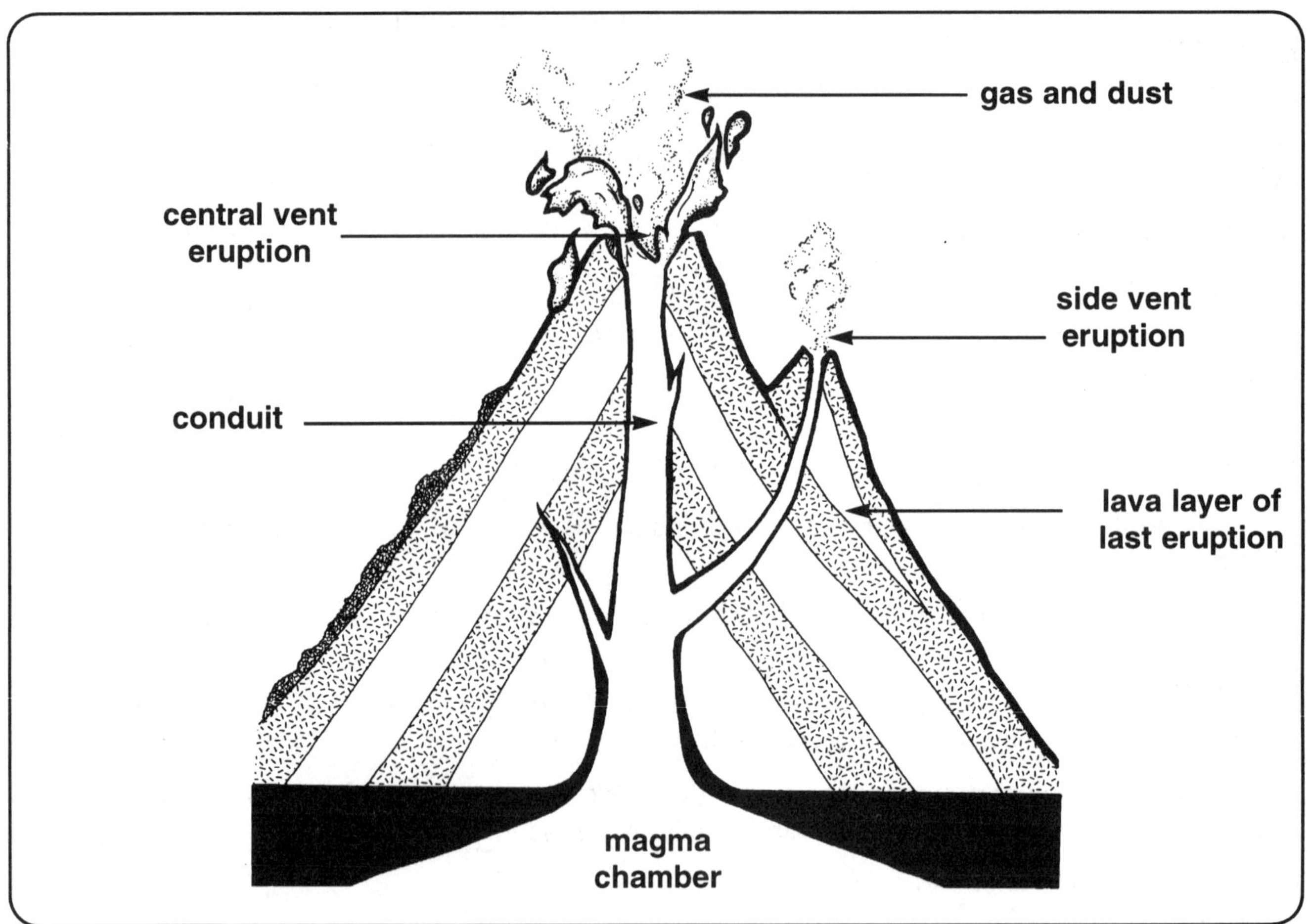

Day 3—Make a Volcano

Materials

- dirt
- baking soda
- vinegar
- liquid dish washing detergent
- small glass jar
- water
- red food coloring

Activity

Make a hill out of dirt, burying the jar within it so that the mouth of the jar is exposed for the volcano's crater.

Put 5 tablespoons (75 mL) of baking soda in the jar. Mix 1 cup (250 mL) of water, 1/2 (125 mL) cup dish washing liquid, and 1/2 cup (125 mL) vinegar in another container. Add food coloring. Pour some into the jar.

"Magma" should bubble up and pour down the sides.

Day 4—Eruptions in Action

Materials

- film or video showing volcanic eruptions, particularly Hawaiian volcanoes

Activity

Show students a film of erupting volcanoes. Tell them that most of the footage available is of the Hawaiian volcanoes. Scientists have generally felt safe studying and filming the Hawaiian volcanoes until recently when there have been more violent and destructive eruptions.

Discuss the film.

Day 5—Review and Reflect

Materials

- "Review and Reflect" form (below)

Activity

Review and discuss concepts introduced during the week. Help students complete the form.

Review and Reflect

Name ______________________________ Date ______________

What I Learned:

Another name for melted rock is ______________________________.

When the rock cools and hardens, it is called ______________________________.

The explosion made by a volcano is called an ______________________________.

The pipe-shaped passage the magma comes through is called a ______________________________.

The opening in the top of a volcano is the ______________________________.

The most interesting thing about volcanoes is ______________________________

____.

Background

Introduce your students to the concepts of erosion and sedimentation as other forces at work to shape the Earth.

Day 1—Three Great Forces

Materials

- reference books and pictures of areas such as the Dust Bowl, the Grand Canyon, and the Mississippi River delta that show erosion and sedimentation

Activity

Say to students:

> We have been studying the changes that come about from inside the Earth. Now we are going to learn about some other forces that shape Earth's surface.
>
> Erosion is the wearing away of soil and rock. The main causes of erosion are moving air (wind), moving water (rainfall, running water, waves), and changes in temperature (freezing and thawing).
>
> Sedimentation is the building up of layers of ground-up rock and soil.
>
> What wears away in one place because of erosion can build up somewhere else as a result of sedimentation.
>
> Let's look at these pictures and talk about what may have caused these landforms to change.

Day 2—Moving Air in Action

Materials

- a sand table filled with a layer of dry soil
- a flat of grass from the nursery
- an electric fan

Activity

Note: Locate this investigation so that you will not have dirt blowing all over your room. Setting up the sand table and flat of grass outside an open window and having the fan in the classroom blowing out through the window might work the best.

Say to students:

> Even though soil that is eroded from one place builds up somewhere else, erosion is a real problem to farmers who are losing their topsoil and can't grow anything anymore. It is important to them to be able to stop or at least slow down the process of erosion.
>
> This sand table full of dry soil is like a field that has not been planted in a place where there has not been enough rain. The flat of grass next to it is like a planted field.
>
> This fan is the wind. Watch what happens when the wind blows. (Turn on the fan.)
>
> Who wants to describe what just happened? Who can guess why it happened?

Give students plenty of time to discuss what they observed and what they concluded.

Day 3—Moving Water in Action

Materials

- a sand table filled with a layer of dry soil
- a flat of grass from the nursery
- a source of water and a hose, if possible (A bucket or watering can will also work.)

Activity

Note: Locate this investigation so that you will not flood your classroom. An outdoor area with tables and benches arranged so everyone can see would be perfect. Set up both the sand table and the flat of grass so they are on a slight slant. Start out with a fine spray of water and gradually increase it to a heavy "rain."

Say to students:

This sand table full of dry soil is a field that has not been planted and has not had any rain. The flat of grass next to it is a planted field.

This water is the rain. Watch what happens during a light rain. (Turn on the water so the dirt is able to absorb the "rain.")

Now watch what happens as the rain continues and becomes heavier.

Who wants to describe what just happened? Who can guess why it happened? (Give students plenty of time to discuss what they observed and what they concluded.)

Day 4—River Deltas

Materials

- a sand table filled with a thick layer of dry soil
- a source of water (a watering can will work well for this)

Activity

Note: It is possible to conduct this investigation inside your classroom without making a mess. It will also take longer than the others, so it will be handier to have it inside. Make a trough the long way in the layer of soil. Either slant the layer of soil or the sand table itself so that your "river" will flow.

Say to students:

This sand table full of dry soil is like a countryside that a river flows through. Water from rain that has fallen into the river at this end runs down the river channel. Let's watch what happens. (Sprinkle water at the high end.)

We will make it "rain" off and on during the day and watch what happens. (The "rain" should start to flow down the river and carry some dirt along with it.)

Who wants to describe what has happened to the river and the land around it? Who can guess why it happened? (Give students plenty of time to discuss what they observed and what they concluded.)

Day 4—River Deltas *(cont.)*

After a river flows for hundreds of years it carries a lot of rich soil with it and forms a delta. This is an area shaped like a triangle at the mouth of the river. The soil there is very fertile. Two very famous deltas are the Nile River delta and the Mississippi River delta.

Note: Save the sand table and tray of grass from the Day 2 activity.

Day 5—Review and Reflect

Materials

- "Review and Reflect" form (See below.)

Activity

Review and discuss concepts introduced during the week. Help students complete the form.

Review and Reflect

Name ______________________________ Date ______________

What I Learned:

The wearing away of rock and soil is called ______________________________.

The building up of layers of ground-up rock and soil is called ______________________.

The two causes of erosion we investigated were ________________________ and

____.

If a farmer wants to protect his soil from erosion, the best thing to do is ______________

____

____

____.

Background

Introduce your students to the idea of composting as a method of building the soil.

Day 1—What Is Soil?

Materials

- a narrow trench dug in a nearby flower bed or other available soil that gets watered regularly
- banana skin
- orange peel
- brown paper bag
- aluminum can
- plastic grocery bag
- Styrofoam packing material
- strip of nylon screen, approximately the length and width of the trench

Activity

Say to students:

What is soil? (the thin layer of material on the Earth's surface that plants can grow in)

What is soil made of? (finely ground-up rock mixed with decayed plants and animal wastes)

How can people make soil better? (by adding things to it that will decay or are biodegradable and that will become part of the soil)

We are going to experiment to see what will decay in the soil and make it better for growing things. I have a banana skin, an orange peel, a brown paper bag, an aluminum can, a plastic grocery bag, and plastic packing material. We are going to bury them. On Friday, we will dig them up and see what has happened.

Bury the items. Place the strip of screen over them before covering them with dirt so that you will be able to lift off the dirt without destroying your experiment.

Day 2—Biodegradable?

Materials

- composting experiment from Day 1
- enough copies of the chart on page 44 for individuals or for groups

Activity

Help students individually or in groups fill out the chart with their predictions about the buried items. Introduce the term "biodegradable" to describe an item that decays and becomes part of the soil.

Save the charts for use in other activities and for Friday when you check your buried items.

Day 2—Biodegradable? *(cont.)*

Biodegradable? Guess and Check

	Guess		Check	
	Yes	No	Yes	No
Banana skin				
Orange peel				
Brown paper bag				
Aluminum can				
Plastic grocery bag				
Packing material				

Day 3—If You Guessed "No"

Materials

- "Biodegradable? Guess and Check" chart from Day 2

Activity

Compare charts to see which items the students guessed would not decay. Write the names of these items on the board.

Brainstorm ideas for dealing with these items. (Some possible answers could be putting them in the trash, taking them to a recycling center, or using them for something else.)

Day 4—Junk Art

Materials

items junk such as the following:

- small broken toys
- pieces from old games
- extra nuts, bolts, and screws
- unidentified small objects, etc.
- square or rectangular wood scraps (a local lumber store might be willing to donate these) for wood base
- low-temperature glue guns
- spray paint (metallic colors work best)

Activity

Under adult supervision, have students glue their collections of junk to a wood base in any pattern they like. Spray the completed art objects with metallic paint and display them.

Day 5—Review and Reflect

Materials

- "Biodegradable? Guess and Check" chart from Day 2
- "Review and Reflect" form (below)

Activity

Dig up the items you buried on Day 1 by carefully lifting the nylon screen.

Fill in the "Check" part of the chart and compare it with the "Guess" part.

Review and discuss concepts introduced during the week. Help students complete the form.

Review and Reflect

Name ________________________________ Date ______________

What I Learned:

Things that decay and turn into soil are ______________________________.

These items decayed: ____

____

____.

These items did not decay: ____

____.

Week 15

Review Lesson

Background

Take this opportunity to review and reinforce the concepts and vocabulary introduced during the preceding four weeks.

Day 1—Earthquakes

Materials

- painted blocks from Week 11, Day 1

Activity

Say to students:

Let's look at these painted blocks and see what we can remember about earthquakes.

Who remembers what a crack in the Earth's surface is called? (fault)

What kinds of movement can occur along a fault? (up, down, sideways)

What should a person do in case of an earthquake? (duck and cover, then leave the building)

What kinds of disaster drills do we have? (answers will vary)

What rules do we follow? (answers will vary)

What things would be useful in case of an emergency? (a flashlight, a radio with fresh batteries, a bottle of water, water purification tablets, nonperishable snacks, a hat or cap with a brim, and a small game or two, all packed in the backpack or something equally portable)

Day 2—Volcanoes

Materials

- a blackboard and chalk

Activity

Say to students:

Let's see what we can remember about volcanoes.

What is another name for melted rock? (magma)

What is liquid rock called after it cools and hardens? (lava)

What is the explosion made by a volcano called? (eruption)

What is the passage the magma comes through called? (conduit)

What is the opening in the top of a volcano called? (crater)

Who would like to come to the board and draw a diagram of the inside of a volcano?

Who would like to label the parts?

Day 3—Erosion

Materials

- sand table full of dirt and the flat of grass from Day 2, week 13

Activity

Ask the students:

Who can tell us what erosion is? (the wearing away of rock and soil)

What causes erosion? (wind, water, temperature changes)

How did we investigate erosion caused by wind? Who can explain?

How did we investigate erosion caused by water? Who can explain?

What did we learn about what a river does? Who can explain?

Discuss all of the concepts.

Day 4—Soil Building

Materials

- charts from Week 14 (See page 44.)

Activity

Ask the students:

What did we learn about soil? (thin layer of material on the Earth's surface that plants can grow in)

What is soil made of? (finely ground-up rock mixed with decayed plants and animal wastes)

How can people make soil better? (by adding things to it that will decay, are biodegradable, and will become part of the soil)

What things did we use for our experiment? (a banana skin, an orange peel, a brown paper bag, an aluminum can, a plastic grocery bag, and plastic packing material)

What happened to the banana skin? Raise your hand if you guessed correctly.

What happened to the orange peel? Raise your hand if you guessed correctly.

What happened to the brown paper bag? Raise your hand if you guessed correctly.

What happened to the aluminum can? Raise your hand if you guessed correctly.

What happened to the plastic grocery bag? Raise your hand if you guessed correctly.

What happened to the plastic packing material? Raise your hand if you guessed correctly.

What do we know now that we didn't know before?

What other uses did we think of for trash or junk?

Day 5—Review and Reflect

Materials

- drawing paper
- pens, pencils, crayons, markers
- "Review and Reflect" form (See below.)

Activity

Have each student draw a picture (diagram) of something that could happen to a river that flowed across a fault line during an earthquake. Accept all versions; they will give you a good idea of each child's conceptual level.

Give students necessary help to complete the form below.

Review and Reflect

Name ______________________________ Date ______________

Choose words from below to complete this story.

Forces inside the Earth make things happen on Earth's ______________.

If rocks near the surface break, they can cause a crack called a ______________ .

Rocks and soil may slip various ways along a fault—up, down, and ______________.

This movement is called an______________________________.

______________________________are also caused by forces inside the Earth.

Hot, melted rock called ______________________________ can force its way to the surface.

The explosion it causes is called an ______________________________.

Forces on the surface of the Earth also make things happen. The wearing away of rocks and soil, called ______________, is caused by ______________, and ______________.

magma	fault	volcanoes
wind	surface	earthquake
sideways	erosion	water
eruption		

Background

Review the rotation and revolution of the Earth. Introduce the idea that Earth's tipped axis causes the seasons.

Day 1—Earth Rotates and Revolves

Materials

- a table lamp that casts light in all directions (not a directional desk lamp)

Activity

Say to students:

Let's review what we learned about the Earth's rotation on its axis and how it revolves around the sun.

Who wants to be the Earth? Remember how the Earth turns around and around (rotates) while it travels around the sun (revolves).

Let as many students as possible play the part of Earth to reinforce these concepts.

Day 2—Earth's Axis Is Tipped

Materials

- a table lamp that casts light in all directions (not a directional desk lamp)
- a globe

Activity

Say to students:

Today we are going to learn something new about the Earth, its axis, and its year-long trip around the sun.

Show the globe to students.

Have you ever noticed how a globe is tipped? It isn't made that way just for fun. It is made that way because the Earth's axis really is tipped. Earth's tipped axis is what causes the seasons.

We will have to do this demonstration with a globe because it is impossible to walk around tipped that way and still rotate.

We will pretend that the distance around the room is the length of the trip that the Earth makes in one year. As I stand here in the back of the room, I am in autumn. To my right is winter. The front of the room is spring. To my left is summer.

The Earth goes around the sun in a counterclockwise orbit. It rotates 365 times in its year-long journey, but its axis always stays tipped the same way. (Start to walk around the room with the globe.)

Day 2—Earth's Axis Is Tipped *(cont.)*

In the winter, Earth's axis is tipped away from the sun. The top half of the globe above the equator, the Northern Hemisphere, gets less heat and light from the sun. (Stop and show the position of the Earth.)

In the spring, the sun is directly over the equator. The Northern Hemisphere and the Southern Hemisphere, the bottom half of the globe below the equator, get equal amounts of light and heat. (Stop and show the position of the Earth.)

In the summer, Earth's axis is tipped toward the sun. The top half of the globe above the equator, the Northern Hemisphere, gets more heat and light from the sun. (Stop and show the position of the Earth.)

In the fall, the sun is directly over the equator again. The Northern Hemisphere and the Southern Hemisphere, the bottom half of the globe below the equator, get equal amounts of light and heat. (Stop and show the position of the Earth.)

The seasons are reversed in the Southern Hemisphere.

Repeat this demonstration as often as possible. Let different students carry the globe and demonstrate its positions.

Day 3—Label the Seasons Diagram

Materials

- individual copies of the diagram below
- a globe

Activity

Help students label the diagram below with the names of the seasons as they occur in the Northern Hemisphere.

Seasons in the Northern Hemisphere

Day 4—Special Days

Materials

- individual copies of the diagram from Day 3
- a globe

Activity

Say to students:

> The days in spring and fall when the hours of daylight and darkness are equal are called equinoxes. There is a spring equinox (around March 21) and a fall equinox (around September 23).
>
> The longest day of the year, the summer solstice, occurs on June 21 or 22. The shortest day of the year, the winter solstice, occurs on December 21 or 22.

Repeat the demonstration with the globe while talking about these special days.

Day 5—Review and Reflect

Materials

- "Review and Reflect" form (below)

Activity

Review and discuss concepts introduced during the week. Help students complete the form.

- -

Review and Reflect

Name ______________________________ Date ______________

What I Learned:

We have four seasons: ________________, ________________,

________________, and ________________.

We have seasons because ______________________________

____.

Background

Introduce the idea of climate as the usual weather in an area over a period of many years.

Day 1—What Is Climate?

Materials

- encyclopedias and other illustrated reference books
- a film or video showing different climate conditions

Activity

Take time to study pictures or view the film.

Say to students:

> Climate is the usual weather in an area over a period of many years. If you say, "We live in an area with hot summers," you are talking about climate. If you say, "It is really hot today," you are talking about weather.
>
> Climate can be classified by temperature (how hot and cold it gets), rainfall (how much it rains), and seasonal variations (when things happen).
>
> Two places might have the same average yearly rainfall, but if the rain falls throughout the year in one place and all during the summer in another place, that is a seasonal variation.

Day 2—Types of Climate

Materials

- encyclopedias and other illustrated reference books
- a film or video showing different climate conditions

Activity

Take time to study pictures or view the film.

Say to students:

> There are nine major types of climate. Three of the most extreme are tropical (hot and wet), desert (hot or cold and dry), and polar (long cold winter, short warm summer, little rain or snow).

Help students to use reference books to find one example of a place with each of these climates:

- tropical
- hot desert
- cold desert
- polar

Day 3—Causes of Different Climates

Materials

- encyclopedias and other illustrated reference books
- a globe
- a classroom map of the world

Activity

Tell students some of the causes for different climates. You may also wish to make a chart of the following terms.

- Latitude: In general, the closer you are to the equator, the hotter the climate.
- Elevation: The higher you are, the colder it gets.
- Surface: Some surfaces hold the heat of the sun more than others.
- Oceans: The oceans do not hold the sun's heat as much as the land does.

Discuss your own area. Which of these factors has an effect on your climate?

Day 4—Effects of Climate

Materials

- encyclopedias and other illustrated reference books
- a globe
- a classroom map of the world

Activity

Discuss these ideas:

How does climate affect clothing? Do you have certain clothes because of the climate you live in?

How does climate affect housing? Do you need a furnace or an air conditioner or both?

How does climate affect transportation? Do you need any special kind of transportation because of your climate?

How does climate affect crops that can be grown in an area? Is the area you live in known for certain kinds of crops?

What effect does climate have on natural plant life? What grows wild in your area?

What effect does climate have on animal life? What wild animals live (or used to live) in your area?

Day 5—Review and Reflect

Materials

- "Review and Reflect" form (below)

Activity

Review and discuss concepts introduced during the week. Help students complete the form.

Review and Reflect

Name ______________________________ Date ______________

What I Learned:

Climate is ______________________________.

Climates can be classified by ______________________________.

Some factors that cause different climates are ______________________________.

Effects of my region's climate on

clothing ______________________________

____

housing ______________________________

____

transportation ______________________________

____

crops ______________________________

____

wild plants ______________________________

____

native animals ______________________________

____

Background

Introduce the idea of weather as the daily variations in climate.

Day 1—Checking the Weather

Materials

- enough newspaper weather pages for individual students or partners
- classroom wall maps of the U.S. and of the world

Activity

This can be an ongoing activity for the week or for as long as you wish to continue it.

Have students check the weather page for cities they are interested in.

One student, the "weatherperson," calls on volunteers one at a time to go to the map and point out a city.

The weatherperson says, "What is the high in _____? What is the low in ___________? What kind of weather are they having?" and calls on various students to find and read those pieces of information from the weather page.

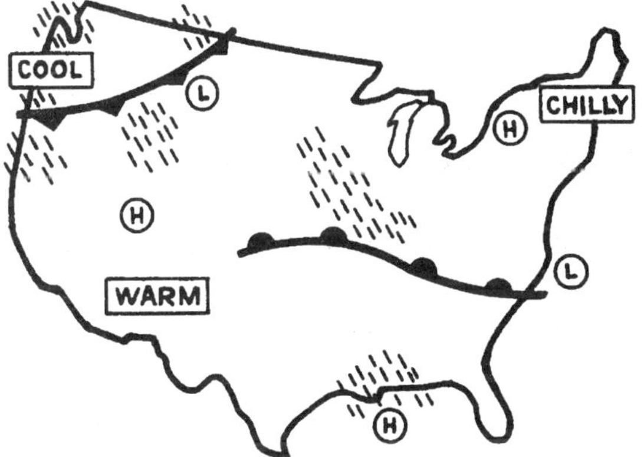

The students at the map can then repeat the name of the city and state (if applicable), country, and continent.

Day 2—Checking Unusual Weather

Materials

- enough newpaper weather pages for individual students or partners
- classroom wall maps of the U.S. and of the world
- calendar or chart (page 56)

Activity

Have students check the weather page for the highest high and the lowest low temperature. (These figures may be readily available, or students may need to look through lists of temperatures to find them.)

Write the names of the two cities and their temperatures on the calendar or chart each day.

Have someone find them on the map.

Day 2—Checking Unusual Weather *(cont.)*

Highest High, Lowest Low					
	Monday	**Tuesday**	**Wednesday**	**Thursday**	**Friday**
High					
City					
Low					
City					

Day 3—Predicting the Weather

Materials

- enough newpaper weather pages for individual students or partners
- classroom wall maps of the U.S. and of the world

Activity

Say to students:

Predictions about the weather, or telling in advance what the weather will be, are called forecasts.

Scientists who study the weather are called meteorologists.

Forecasts are important to people who want to plan activities or decide what to wear.

Forecasts are even more important to farmers who need to protect their crops from frost or ranchers who need to shelter their cattle or sheep from storms.

Forecasts are vital to the people who schedule the space program.

Forecasts can be a matter of life or death to people who are threatened with hurricanes or floods.

Even though weather forecasting is so important, it is not an exact science and the forecasts cannot be made for very long ahead with any accuracy because the weather is always changing.

The map shown on the weather page of the daily newspaper shows how meteorologists look at the weather by plotting the cold fronts and warm fronts and the highs and the lows.

Day 4—How Weather Is Made

Materials

- enough newpaper weather pages for individual students or partners
- classroom wall maps of the U.S. and of the world

Activity

Tell students that the elements that make up the weather are

- temperature: people measure how high or low it is with a thermometer.
- wind: the important things to know about wind are direction and speed.
- moisture: moisture can fall out of the air as rain or snow (called precipitation) or stay in the air as humidity.
- air pressure: since warm air rises, it creates an area of low pressure called a low. Cold air is heavier. It creates an area called a high.

Have students find highs and lows on the newspaper weather map.

Day 5—Review and Reflect

Materials

- "Review and Reflect" form (below)

Activity

Review and discuss concepts introduced during the week. Help students complete the form. Depending on the age group you are teaching, you may want students to dictate their answers to you or to an aide.

Review and Reflect

Name ______________________________ Date ______________

What I Learned:

Weather is ______________________________.

Scientists who study the weather are called ______________________________.

Weather predictions are called ______________________________.

It is important to know what the weather will be because ______________________________

______________________________.

Background

Introduce the concept of the water cycle.

Day 1—Evaporation

Materials

- chart of the water cycle
- several shallow dishes or pans
- "Evaporation Chart" (below)
- measuring cup
- water

Activity

Show students the chart of the water cycle and share some facts. You may wish to reproduce the following information for each student or write it on a chart or overhead transparency.

The Earth has always had the same amount of water that is has today.

Earth's water is constantly being renewed by going through the water cycle.

First, water evaporates into the air off the surfaces of oceans, lakes, ponds, and so on.

Then the water vapor in the air condenses and gathers into clouds.

When there is too much moisture for the clouds to hold, the water falls (precipitates) as rain or snow.

The water then flows into streams and rivers, which run back into oceans and lakes or sink deep into the ground. It takes a long time for rain or snow to get back to a body of water. But whether it does so quickly or slowly, moisture will evaporate again from somewhere and start the cycle all over again.

Say to students:

> Today we are going to create our own bodies of water and see how long it takes them to evaporate. We will put the same amount of water in each pan. We can put one where it is warm and another where it will stay cool. We can let the fan blow on another one. We will keep a record of what happens.

Evaporation Chart

We put one cup (250 mL) of water in each pan. Here is what happened in each pan.

Pan in warm place ____

Pan in cool place ____

Pan near fan ____

Day 2—Clouds

Materials

- chart of the water cycle
- pictures of the three main types of clouds: cirrus, stratus, cumulus

Activity

The second step in the water cycle is when water vapor in the air condenses and forms clouds. Study and discuss pictures of the three main types of clouds. Have students draw and label the three types of clouds.

Clouds

________________ ________________ ________________

Day 3—Rain and Snow

Materials

- chart of the water cycle
- paper and art materials

Activity

The third step in the water cycle takes place when the water in the clouds precipitates or falls to the ground in the form of rain or snow.

Have students draw pictures of rain or snowy-day scenes.

Day 4—Return of the Water

Materials

- chart of the water cycle

Activity

The fourth and last step in the water cycle takes place when the water returns to the oceans. Talk about some of the different ways this happens. You may wish to reproduce the following information for students individually or on a chart or overhead transparency.

Water runs down slopes and forms streams and rivers that flow into lakes or back to the ocean.

Water falls on fields where the plants take it in and breathe it back into the air in the form of water vapor. It will fall again as rain and start another trip back to the ocean.

Water sinks down deep into the ground and comes up through wells or springs.

Day 5—Review and Reflect

Materials

- "Review and Reflect" form (below)

Activity

Review and discuss concepts introduced during the week. Help students complete the form.

Review and Reflect

Name ______________________________ Date ______________

What I Learned:

Earth's water makes itself new again by going through the ______________________.

The four steps in the water cycle are as follows:

1. evaporation, when the water ______________________________
2. condensation, when the water ______________________________
3. precipitation, when the water ______________________________
4. run-off, when the water ______________________________

Week 20

Review Lesson

Background

Take this opportunity to review and reinforce the concepts and vocabulary introduced during the preceding four weeks.

Day 1—Seasons

Materials

- a globe

Activity

Say to students:

Who remembers what causes Earth's seasons? (Earth's axis is tipped.)

Who would like to do the demonstration we did when we studied the seasons?

First, pretend that the distance around the room is a year. The back of the room is autumn. To the right is winter. The front of the room is spring. To the left is summer.

The Earth goes around the sun in a counterclockwise orbit. It rotates 365 times in its year-long journey, but its axis always stays tipped the same way. (Have student start to walk around the room with the globe.)

In the winter, Earth's axis is tipped away from the sun. The top half of the globe above the equator, the Northern Hemisphere, gets less heat and light from the sun. (Have student stop and show the position of the Earth.)

In the spring, the sun is directly over the equator. The Northern Hemisphere and the Southern Hemisphere, the bottom half of the globe below the equator, get equal amounts of light and heat. (Have student stop and show the position of the Earth.)

In the summer, Earth's axis is tipped toward the sun. The top half of the globe above the equator, The Northern Hemisphere, gets more heat and light from the sun. (Have student stop and show the position of the Earth.)

In the fall, the sun is directly over the equator again. The Northern Hemisphere and the Southern Hemisphere, the bottom half of the globe below the equator, get equal amounts of light and heat. (Have student stop and show the position of the Earth.)

What happens in the Southern Hemisphere while this is going on? (The seasons are reversed in the Southern Hemisphere.)

What are the days called when the hours of daylight and darkness are equal? (equinoxes)

What is the longest day of the year called? (the summer solstice)

What is the shortest day of the year called? (the winter solstice)

Repeat the demonstration with the globe while talking about these special days.

Day 2—Climate

Materials

- none

Activity

Say to students:

Who remembers what climate is? (Climate is the general weather in an area over a period of many years.)

What are you talking about when you say, "We live in an area with hot summers"? (climate)

What are you talking about when you say, "It is really hot today"? (weather)

When you talk about climate, what does temperature mean? (how hot and cold it gets)

What does rainfall mean? (how much it rains)

What kind of weather is tropical? (hot and wet) Desert? (hot or cold and dry) Polar? (long cold winter, short warm summer, little rain or snow)

What are some of the causes for different climates? (latitude, elevation, surface features, and oceans)

What are some of the things climate affects? (clothing, housing, transportation, crops that can be grown, natural plant life, and native animals)

Discuss these concepts as necessary for review.

Day 3—Weather

Materials

- newspaper weather pages, if desired

Say to students:

Who remembers what predictions about the weather are called? (forecasts)

What are scientists who study the weather called? (meteorologists)

Who might weather forecasts be important to? (people who want to plan activities or decide what to wear, farmers who need to protect their crops from frost or ranchers who need to shelter their cattle or sheep from storms, the people who schedule the space program, and people who are threatened with hurricanes or floods)

What are the elements that make up the weather? (temperature, wind direction and speed, moisture, and air pressure)

Day 4—Water Cycle

Materials

- chart of the water cycle

Activity

Say to students:

Let's look at the chart of the water cycle and review what we have learned.

How does water get into the air in the first place? (evaporation)

What did we learn when we set out our pans of water? (Refer to "Evaporation Chart," page 58.)

After water evaporates into the air, what happens? (It condenses and gathers into clouds.)

What happens to the clouds? (When they contain too much moisture to hold, it falls in the form of rain or snow.)

After the moisture falls, what happens? (The water makes its way back to the oceans.)

Day 5—Review and Reflect

Materials

- drawing paper
- pens, pencils, crayons, markers
- "Review and Reflect" form (below)

Activity

Have each student draw a picture (diagram) of the Earth at each season of the year.

Have each student draw a picture (diagram) of the water cycle.

These student-generated responses will give you a real look at each child's concept level.

Give students necessary help to complete the form below. Depending on the age group you are teaching, they can do it themselves or dictate their information to you or to an aide.

Review and Reflect

Name ______________________________ Date ______________

We need to take care of the Earth's water because ______________________________

____.

Background

Introduce geology as a science that uses the Earth itself as a history book.

Day 1—The Earth as a History Book

Materials

- encyclopedias and other illustrated reference books

Activity

Say to students:

Geology is the study of the Earth in which the Earth itself it used as a history book.

The Earth is and has been always changing. Geologists find out how it has changed by looking at the layers piled one on top of another.

In places where a river has cut through or a mountain has pushed up, we can see the layers and "read" the meaning.

A layer of lava means that there was once an eruption from a volcano.

A layer of coal shows where there was once a swamp.

A layer of a certain kind of rocks shows a place that a glacier moved through.

A layer filled with petrified shells means the area was once covered by an ocean or sea.

Day 2—Make a Model

Materials

- encyclopedias and other illustrated reference books
- a small empty aquarium or terrarium
- pieces of lava
- pieces of coal (charcoal will work)
- some gravel
- sand and small shells
- soil
- poster board and marking pens

Activity

Have students build layers of the materials you have gathered in the aquarium or terrarium.

Make certain the layers are thick enough to be easily distinguished from one another.

Stand the poster board next to the display and use marking pens to label the layers. (See page 65.)

Call your display "Layers of Earth's History."

Day 2—Make a Model *(cont.)*

Layers of Earth's History

Most Recent

Soil

Gravel

Coal

Lava

Sand and Small Shells

Oldest

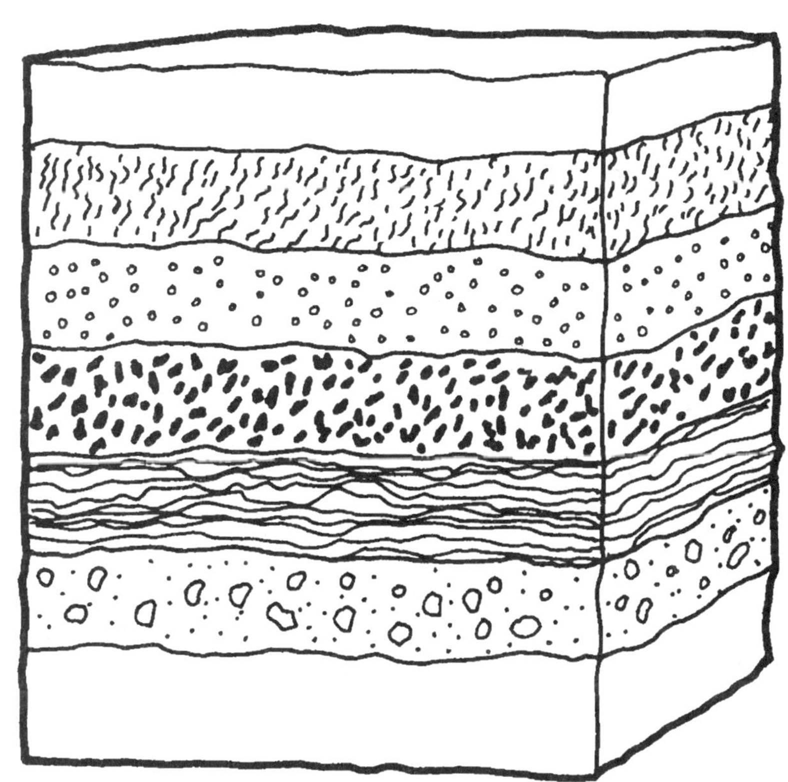

Day 3—Telling the Story of Earth's Layers

Materials

- writing materials

Activity

Have students look at the model they have constructed and, starting at the bottom, tell what they think happened to the Earth based on the layers represented in the model.

Depending on the age group you are teaching, you may want to give students the option of dictating their stories to you or to an aide.

Day 4—Geologists

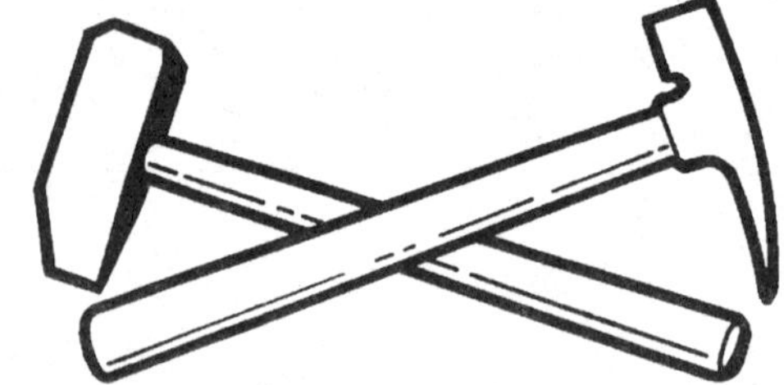

Materials

- writing materials

Activity

Say to students:

At one time, being a geologist meant being a scientist who studied everything about the Earth.

Now the subject of studying the Earth has gotten so big that geology has split into many different specialties.

Geologists today study the following:

- fossils
- volcanoes
- rocks and minerals
- earthquakes
- mountains
- glaciers
- petroleum
- ground water

Each of these specialties has its own name.

Ask students:

If you were going to be a geologist, which kind would you want to be? Why?

Give students a chance to express and discuss their choices.

Day 5—Review and Reflect

Materials

- "Review and Reflect" form (See below.)

Activity

Review and discuss concepts introduced during the week. Help students complete the form. Depending on the age group you are teaching, you may want students to dictate their answers to you or to an aide.

Review and Reflect

Name ______________________________ Date ______________

Why do geologists want to know about what happened to the Earth in the past? ____________

____

____.

Background

Give your students some interesting geological information about Earth's mountains.

Day 1—Kinds of Mountains

Materials

- study prints or travel videos of different mountains
- "Kinds of Mountains" (See below.)

Activity

Say to students:

There are four main types of mountains: folded mountains, fault-block mountains, dome mountains, and volcanic mountains.

Folded mountains occur when the Earth's crust makes great waves.

Fault-block mountains happen when the Earth's crust breaks into great blocks; some move up and some down, like the blocks we used when we talked about earthquakes.

Dome mountains occur when the Earth's crust does not break or fold but simply rises in great round domes something like huge blisters.

Volcanic mountains form when a volcano erupts over and over again, gradually building huge peaks.

Have students use the chart below to draw their impressions of these different kinds of mountains.

Kinds of Mountains

Folded Mountains	**Fault-block Mountains**
Dome Mountains	**Volcanic Mountains**

Day 2—Great Mountain Ranges

Materials

- study prints of different mountains or a video showing mountain regions (Travel videos are good choices.)
- classroom wall map of the world
- individual enlarged copies of map (See below.)

Activity

Have students compare the mountain ranges shown on their maps to the classroom world map and identify the mountain ranges on the large world map.

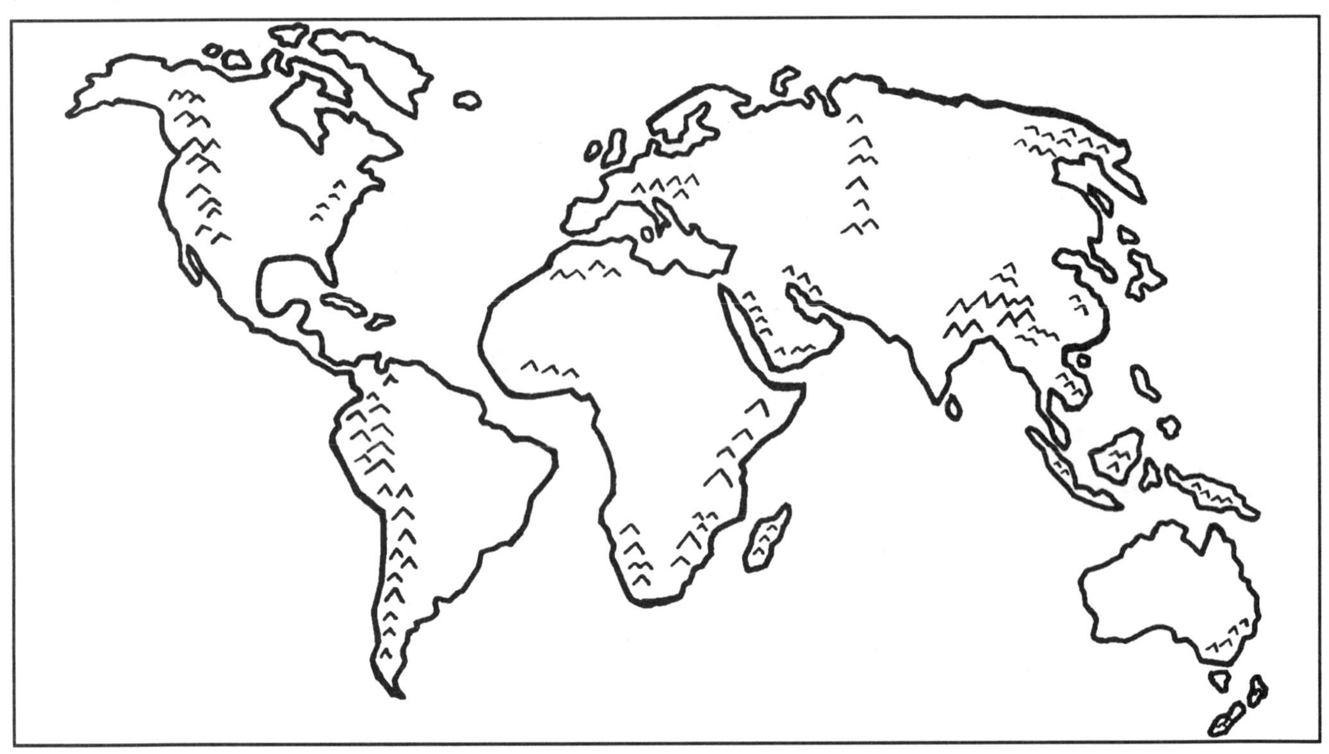

Day 3—Famous Mountains

Materials

- encyclopedias and other illustrated reference materials

Activity

Help students use reference materials to discover one fact about each of these famous mountains.

- Mt. Everest
- Mauna Loa
- Kilimanjaro
- Mt. Fuji
- Matterhorn

Day 4—Mountain Climbing

Materials

- encyclopedias and other illustrated reference materials

Activity

Ask students what equipment they think they would need to take on a mountain climbing expedition.

List suggestions on the board and discuss them.

If no one suggests oxygen equipment, make the suggestion yourself and discuss the concept that the air contains less oxygen at high altitudes. (Maybe some of the students have been on airplanes where the use of oxygen equipment was explained. Encourage them to share their information.)

Have students discuss possible reasons why people want to climb high mountains.

Day 5—Review and Reflect

Materials

- "Review and Reflect" form (See below.)

Activity

Review and discuss concepts introduced during the week. Help students complete the form. Depending on the age group you are teaching, you may want students to dictate their answers to you or to an aide.

Review and Reflect

Name ______________________________ Date ______________

What I Learned:

The four main kinds of mountains are ______________________________

______________________________, ______________________________,

and ______________________________.

The eight leading mountain ranges are ______________________________

____

______________________________.

Background

Relate the glaciers of today to the Ice Age, prehistoric animals, and prehistoric people.

Day 1—What Is a Glacier?

Materials

- encyclopedias and other illustrated reference materials
- pictures or a video of existing glaciers as found in Alaska, Canada, or Antarctica

Activity

Look at the pictures or video and discuss what you saw.

Say to students:

> A glacier is a huge river of ice. Glaciers move just like regular rivers but only a few inches a day. Some glaciers move toward an ocean. When they reach the water, pieces break off the glacier and form huge icebergs. Some glaciers flow down mountains and through valleys, feeding mountain lakes.

Day 2—The Ice Age

Materials

- encyclopedias and other illustrated reference materials
- classroom wall map of the world
- a globe
- large shallow cardboard box (See page 71.)
- construction paper and assorted art materials

Activity

Use the map and globe to point to appropriate areas as you *say to students*:

> The Ice Age began about 1,000,000 years ago and lasted until a time between 10,000 and 25,000 years ago.
>
> During the Ice Age, great sheets of ice called "continental glaciers" covered much of the northern part of the Earth. These glaciers moved south and then moved back (melted) four times during that time.
>
> When glaciers moved south they gouged out deep places in the Earth. When they moved back, some of these places filled up with water. This is how the Great Lakes were formed.
>
> Some of the places gouged out by glaciers became deep valleys. This is how Yosemite Valley was formed.
>
> So much water was trapped in these sheets of ice that the oceans were 300 feet lower than they are today. When the glaciers melted, the oceans filled up to their present level.

Day 2—The Ice Age *(cont.)*

Help students design the background for an Ice Age diorama, using the inside surface of a cardboard box to portray a scene with mountains and glaciers. The bottom surface or "ground" can be covered with soil and rocks.

Day 3—Animals of the Ice Age

Materials

- encyclopedias and other illustrated reference materials
- classroom wall map of the world or globe
- Ice Age diorama background from Day 2
- small plastic figures of Ice Age animals such as wooly mammoths, if possible

Activity

Tell students that many animals lived during the Ice Age. Some of them, like the horse, camel, and elephants, are still in existence. Others, such as wooly mammoths, cave bears, and giant ground sloths, have become extinct.

Have students add figures of these animals to their diorama. Use small plastic figures or help students research and draw small replicas of the animals, mount them on stiff paper, glue a flap on the back to help them stand up, and add them to the diorama.

Day 4—People of the Ice Age

Materials

- encyclopedias and other illustrated reference materials
- classroom wall map of the world
- globe
- Ice Age diorama background from Day 2 and Day 3

Activity

Say to students:

> The Ice Age and the Old Stone Age—the first part of the Stone Age—happened at about the same time.
>
> The Stone Age is so called because people used stone tools. We know about these people because scientists have been able to find and date these tools.
>
> We also know about these people because they made beautiful paintings on the walls of caves.
>
> The "cave men" (and women, too, of course) used fire, wore clothing made of animal skins, and got their food by hunting and gathering.

Have students add figures of people to complete their diorama of the Ice Age.

Day 5—Review and Reflect

Materials

- "Review and Reflect" form (See below.)

Activity

Review and discuss concepts introduced during the week. Help students complete the form.

Review and Reflect

Name ______________________________ Date ______________

What I Learned:

Glaciers are ____.

During the Ice Age, ______________________ covered the northern part of the Earth.

Two animals that lived during the Ice Age were ____________ and ____________.

We know about Stone Age people from their ____________ and ____________.

Background

Rocks are another area of interest to geologists.

Day 1—Kinds of Rocks

Materials

- encyclopedias and other illustrated reference materials
- samples of the three main types of rocks: igneous, sedimentary, metamorphic (Check your district for rock kits or buy inexpensive rock samples at a hobby store.)

Activity

Pass the rocks around for students to look at and feel.

Say to students:

There are three main kinds of rocks, and they have long, interesting names:

Igneous rocks are formed by heat from very hot matter (magma) that has cooled and hardened. Lava is an example of igneous rock. (We learned about this when we talked about the layers inside the Earth and about volcanoes.)

Sedimentary rocks are formed by pressure. When wind and water wear away pieces of older rocks, they are washed away and deposited in layers. Over millions of years, the pressure of all the top layers pushing down on the bottom layers turns them to rock. Sandstone is a kind of sedimentary rock.

Metamorphic rocks are rocks that have changed because of either heat or pressure. Marble is a metamorphic rock.

Ask students who have rock collections to bring them in to share on Day 2.

Day 2—Rock Collections

Materials

- students' rock collections
- construction paper
- marking pens

Activity

Lay out sheets of construction paper on counters and shelves and ask students to display their collections on the paper. They can write their names on the paper where their collections are arranged.

Students may want to briefly introduce their collections to the class. They can tell where they collected their rocks, how they got started collecting them, and so on.

Allow time for everyone to look at, enjoy, and ask questions about all the collections and samples.

Day 3—Identifying Rocks

Materials

- an inexpensive reference collection (These can be purchased from hobby stores or rock and mineral dealers.)
- students' rock collections
- a selection of rocks for those who do not have collections
- individual student copies of "Rock Record" (See below.)

Activity

Encourage students to try to identify their rock samples by comparing them with the reference collection.

Show students how to paint a small white dot on each sample and, when the paint has dried, give each sample a number. They can then keep track of information about each of their rocks.

Rock Record

Specimen #	Date Found	Place Found	Kind of Rock

Day 4—Other Resources for Rocks

Materials

- brochures and information about local museums
- an expert or hobbyist to talk to your class, if possible

Activity

Say to students:

Most museums have collections of rocks and minerals that are fun to see.

Distribute the materials about local displays and let interested students either borrow brochures to take home or copy down relevant information to share with their parents.

If you can find a local expert or "rock hound" to come in and talk to your class, this would be an interesting addition to your science curriculum.

Day 5—Review and Reflect

Materials

- "Review and Reflect" form (See below.)

Activity

Review and discuss concepts introduced during the week. Help students complete the form.

Review and Reflect

Name ______________________________ Date ______________

What I Learned:

Igneous rocks are formed by ______________________________.

Sedimentary rocks are formed by ______________________________.

Metamorphic rocks have been changed by either ______________ or ______________.

Lava is an ______________________________ rock.

Sandstone is a ______________________________ rock.

Marble is a ______________________________ rock.

Rock collecting is an interesting hobby because ______________________________.

Background

Take this opportunity to review and reinforce the concepts and vocabulary introduced during the preceding four weeks.

Day 1—Geology

Materials

- "Layers of Earth's History" model from Week 21

Activity

Say to students:

Who remembers how geologists "read" the history of the Earth?

What does a layer of lava mean? (that there was an eruption from a volcano)

What does a layer of coal mean? (that there was a swamp)

How can geologists tell if a glacier moved through? (by looking at certain kinds of rocks)

What does a layer filled with petrified shells mean? (that the area was once covered by an ocean or sea)

Geologists divide up their study of the Earth into different specialties. Who can name some of them? (fossils, rocks and minerals, mountains, volcanoes, earthquakes, glaciers, petroleum, ground water)

Take time to answer questions and discuss the concepts.

Day 2—Mountains

Materials

- classroom wall map of the world
- map of mountain ranges from Week 22

Activity

Say to students:

Who remembers one of the four main types of mountains? (folded mountains, fault-block mountains, dome mountains, and volcanic mountains)

Who can name one of the Earth's main mountain ranges and show us where it is on the map? (refer to the map of mountain ranges from Week 22)

Who can name one of the famous mountains we looked up and tell us one fact about it? (Mt. Everest, Mauna Loa, Kilimanjaro, Mt. Fuji, Matterhorn)

Who can show us those mountains on the map?

Take time to let everyone participate.

Day 3—Ice Age and Glaciers

Materials

- encyclopedias and other illustrated reference materials
- Ice Age dioramas (See page 71.)

Activity

Ask students:

What is a glacier? (a huge river of ice)

Where are glaciers found today? (Alaska, Canada, Antarctica)

What was the Ice Age? (a period of time during which great sheets of ice called continental glaciers covered much of the northern part of the Earth—these glaciers moved south and then moved north four times during that period)

What are some famous landmarks that were carved out by the glaciers? (the Great Lakes, Yosemite Valley)

What happened to the oceans during the Ice Age? (They were 300 feet/90 meters lower than they are today.)

Can you name some of the animals that lived during the Ice Age? (the horse, camel, and elephant, which are still in existence, and the wooly mammoth, cave bear, and giant ground sloth, which have become extinct)

What is another name for the period called the Ice Age? (the Old Stone Age)

Why was it given that name? (because people used stone tools)

Besides their tools, what are the people of the Stone Age known for? (beautiful paintings on the walls of caves)

What are these people sometimes called? (cave men or cave people)

What are some other things we know about these people? (that they used fire, wore clothing made of animal skins, and got their food by hunting and gathering)

Day 4—Rocks and Minerals

Materials

- samples of the three main types of rocks: igneous, sedimentary, metamorphic (Check your district for rock kits or buy inexpensive rock samples at a hobby store.)

Activity

Pass the rocks around for students to look at and feel.

Say to students:

Look at these rocks and see how much you can remember of what we discussed.

What kind of rock is this? (indicate the igneous rock)

Day 4—Rocks and Minerals *(cont.)*

How are igneous rocks formed? (by heat from very hot matter [magma] that has cooled and hardened)

Which kind of rock is this? (indicate the sedimentary rock)

How are sedimentary rocks formed? (by pressure)

Does anyone remember how this happens? (When wind and water wear away pieces of older rocks, they are then washed away and deposited in layers. Over millions of years the pressure of all the layers pushing down turns the bottom layers to rocks.)

What are metamorphic rocks? (rocks that have changed because of either heat or pressure)

Day 5—Review and Reflect

Materials

- drawing paper
- pens, pencils, crayons, markers
- "Review and Reflect" form (See below.)

Activity

This "Review and Reflect" is a student-generated response. Give any necessary help in reading and understanding the directions but try not to help with the actual answers.

These student-generated responses will give you a look at each child's conceptual level. Check, for example, to see if the most recent layer is on the top of the diagram and the oldest on the bottom.

Review and Reflect

On another piece of paper draw a picture (diagram) showing the layers of a part of the Earth where the following events happened:

First, the area was covered by a shallow inland sea.

Second, the area turned into a swamp.

Third, volcanoes erupted and covered the area.

Fourth, glaciers moved back and forth across the land.

Fifth, the area looked the way it does now.

Be sure to label the different layers.

Background

Paleontology is a career that appeals to young children with its emphasis on camping, digging in the dirt, looking for clues, finding treasures such as dinosaur bones, and then fitting the pieces of bone together like a jigsaw puzzle.

Day 1—Scientists Who Have Fun

Materials

- encyclopedias and other illustrated reference materials
- *Digging Up Dinosaurs* by Aliki
  (Let's Read Find-Out Science Book, HarperCollins Children's Books, 1988)

Activity

Say to students:

This week we will be learning about paleontology. Paleontology deals with the study of fossils. Fossils of plants and animals are found in layers of rock like the ones we have been studying, and we will learn more about them later in the week.

Scientists who study fossils are called paleontologists.

While many scientists work all day in labs and offices, paleontologists work "out in the field" with other people who are interested in the same things, spending months at a time camping in wild places. They hire people to help them dig in the dirt to look for fossils. They use picks and shovels and even toothbrushes to find things in the Earth.

Paleontologists look for clues that will lead them to treasures. The treasures that they find are often the bones of dinosaurs.

After they find the bones and dig them up, paleontologists go back to their museums and fit the bones together like giant jigsaw puzzles.

This book, *Digging Up Dinosaurs* by Aliki, tells about these scientists.

Read the book aloud and discuss it with students.

Day 2—Different Kinds of Fossils

Materials

- encyclopedias and other illustrated reference materials
- sample fossils (Check your school district or nearby natural history museum to borrow samples or purchase a few inexpensive ones.)

Activity

Say to students:

Fossils are records or remains of plants and animals that lived in the past. They are found in layers of rocks, usually sedimentary rock. These plants and animals were alive at the time the rock layer formed.

Day 2—Different Kinds of Fossils *(cont.)*

There are four main kinds of fossils.

Petrified fossils are remains of animals and plants that have actually turned to stone because minerals in the ground water have replaced the original material. (Some students may have visited the Petrified Forest where there are many trees that turned to stone in this way.)

Natural molds are formed when something living is buried in a material that hardens around it. When the body decays and dissolves, it leaves a hole that is the same shape that the body was originally. We are going to make a model of one of these types of fossils tomorrow.

Prints are the actual footprints of extinct animals. Conditions had to be just right for these tracks to be preserved, but there are quite a few of them. You can see them in some museums.

Whole animals and plants are rarely found. Those that have been found were preserved when the animals or plants fell into the tar pits or were trapped in deep cracks in ice.

Pass around any fossils you were able to obtain for the students to hold and examine.

Day 3—Making a Mold

Materials

- modeling clay
- paper towels (several for each student)
- tool to flatten clay (A primary pencil will work.)
- ice cubes

Activity

Say to students:

Today we are going to make molds that work much like natural fossils mold.

Work with a partner to roll out two sheets of clay about 6" x 6" (15 cm x 15 cm).

Use the point of your pencil to make a couple of holes in one of the sheets of clay.

Place an ice cube in the middle of this sheet of clay. Put the other sheet on top and press it down around the ice cube, sealing the clay tight.

Set the clay down on your paper towels gently so that you don't press on the holes you made and close them up. Tomorrow, we will see what happened.

Day 4—Back at the Museum

Materials

- encyclopedias and other illustrated reference materials
- clay molds from Day 3
- safe knives (plastic picnic knives or pumpkin-carving knives)
- several small plastic model kits of dinosaurs (enough for your cooperative groups to each construct one)

Activity

Say to students:

Use your knife to carefully cut around the mold at the bottom of the ice cube. Don't squeeze because, unlike a natural mold, your molds have not had time to harden.

Turn the mold over to look inside. The ice has melted. The water has run out through the holes, leaving behind an impression of the ice cube. If we filled the molds up with water and put them in the freezer, we would have ice cubes like the original ones. That is very much the way a scientist reconstructs a fossil from a natural mold.

Now we are going to do something else that paleontologists do when they are back at the museum. If they were lucky enough to find fossil dinosaur bones, they figure out how to put them together. We are going to pretend that we found the parts in these dinosaur kits, and each group will have a dinosaur to put together. You may start today and continue when you have time during the next three weeks.

Day 5—Review and Reflect

Materials

- "Review and Reflect" form (See below.)

Activity

Review and discuss concepts introduced during the week. Help students complete the form.

Review and Reflect

Name ______________________________ Date ________________

What I Learned:

Scientists who study fossils are called ______________________________.

The four main kinds of fossils are ________________ , ________________,

________________ and ______________________________.

Background

Plan to have students put together individual dinosaur books over the course of the next few weeks.

Day 1—Apatosaurus

Materials

- encyclopedias and other current illustrated reference materials
- access to school library, if available
- enlarged copies of the dinosaur illustrations on page 84 for each student
- writing materials
- drawing and coloring materials

Activity

Say to students:

Color and cut out the picture of Apatosaurus.

Use reference books to get information about Apatosaurus and write a paragraph about this dinosaur. (Depending on the age group you are teaching, you may want to read the information aloud to students and have them dictate their paragraphs to you.)

Mount your colored picture at the top of a piece of paper. Copy your paragraph under the picture. Be sure your name is on your paper before you hand it in.

Note: Apatosaurus used to be called Brontosaurus. Depending on how old your reference books are, you may need to look for information under that name.

Day 2—Stegosaurus

Materials

- access to school library, if available
- writing materials
- drawing and coloring materials
- enlarged copies of the dinosaur illustrations on page 84 for each student
- encyclopedias and other current illustrated reference materials

Activity

Say to students:

Color and cut out the picture of Stegosaurus.

Use reference books to get information about Stegosaurus and write a paragraph about this dinosaur. (Depending on the age group you are teaching, you may want to read the information aloud to students and have them dictate their paragraphs to you.)

Mount your colored picture at the top of a piece of paper. Copy your paragraph under the picture. Be sure your name is on your paper before you hand it in.

Day 3—Ankylosaurus

Materials

- encyclopedias and other current illustrated reference materials
- access to school library, if available
- enlarged copies of the dinosaur illustrations on page 84 for each student
- writing materials
- drawing and coloring materials

Activity

Say to students:

Color and cut out the picture of Ankylosaurus.

Use reference books to get information about Ankylosaurus and write a paragraph about this dinosaur. (Depending on the age group you are teaching, you may want to read the information aloud to students and have them dictate their paragraphs to you.)

Mount your colored picture at the top of a piece of paper. Copy your paragraph under the picture. Be sure your name is on your paper before you hand it in.

Day 4—Triceratops

Materials

- encyclopedias and other current illustrated reference materials
- access to school library, if available
- enlarged copies of the dinosaur illustrations on page 84 for each student
- writing materials
- drawing and coloring materials

Activity

Say to students:

Color and cut out the picture of Triceratops.

Use reference books to get information about Triceratops and write a paragraph about this dinosaur. (Depending on the age group you are teaching, you may want to read the information aloud to students and have them dictate their paragraphs to you.)

Mount your colored picture at the top of a piece of paper. Copy your paragraph under the picture. Be sure your name is on your paper before you hand it in.

Day 5—Review and Reflect

Materials

- "Review and Reflect" form (See below.)

Activity

Review and discuss concepts introduced during the week. Help students complete the form. Depending on the age group you are teaching, you may want students to dictate their answers to you or to an aide.

Review and Reflect

Name ______________________________ Date ______________

What did all of the dinosaurs you studied this week have in common?

____

____

____

____

Illustrations for Week 27

Apatosaurus

Stegosaurus

Ankylosaurus

Triceratops

Background

Continue to put together individual dinosaur books over the course of this unit.

Day 1—Allosaurus

Materials

- encyclopedias and other current illustrated reference materials
- access to school library, if available
- enlarged copies of the dinosaur illustrations on page 87 for each student
- writing materials
- drawing and coloring materials

Activity

Say to students:

Color and cut out the picture of Allosaurus.

Use reference books to get information about Allosaurus and write a paragraph about this dinosaur. (Depending on the age group you are teaching, you may want to read the information aloud to students and have them dictate their paragraphs to you.)

Mount your colored picture at the top of a piece of paper. Copy your paragraph under the picture. Be sure your name is on your paper before you hand it in.

Day 2—Tyrannosaurus

Materials

- encyclopedias and other current illustrated reference materials
- access to school library, if available
- enlarged copies of the dinosaur illustrations on page 87 for each student
- writing materials
- drawing and coloring materials

Activity

Say to students:

Color and cut out the picture of Tyrannosaurus.

Use reference books to get information about Tyrannosaurus and write a paragraph about this dinosaur. (Depending on the age group you are teaching, you may want to read the information aloud to students and have them dictate their paragraphs to you.)

Mount your colored picture at the top of a piece of paper. Copy your paragraph under the picture. Be sure your name is on your paper before you hand it in.

Day 3—Compsognathus

Materials

- enlarged copies of the dinosaur illustrations on page 87 for each student
- encyclopedias and other current illustrated reference materials
- access to school library, if available
- writing materials
- drawing and coloring materials

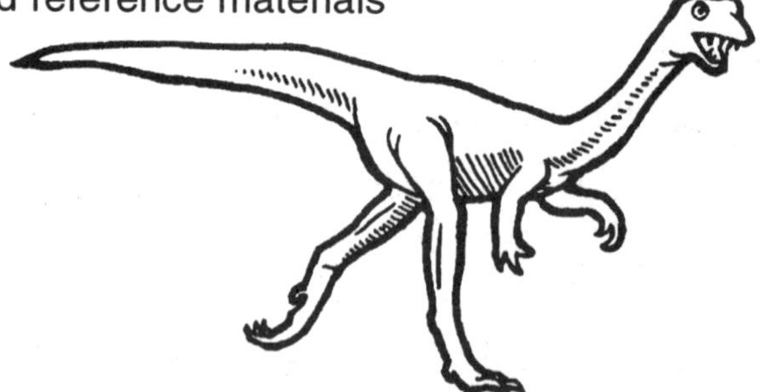

Activity

Say to students:

Color and cut out the picture of Compsognathus.

Use reference books to get information about Compsognathus and write a paragraph about this dinosaur. (Depending on the age group you are teaching, you may want to read the information aloud to students and have them dictate their paragraphs to you.)

Mount your colored picture at the top of a piece of paper. Copy your paragraph under the picture. Be sure your name is on your paper before you hand it in.

Day 4—Pteranodon

Materials

- enlarged copies of the dinosaur illustrations on page 87 for each student
- encyclopedias and other current illustrated reference materials
- access to school library, if available
- writing materials
- drawing and coloring materials

Activity

Say to students:

Color and cut out the picture of Pteranodon.

Use reference books to get information about Pteranodon and write a paragraph about this dinosaur. (Depending on the age group you are teaching, you may want to read the information aloud to students and have them dictate their paragraphs to you.)

Mount your colored picture at the top of a piece of paper. Copy your paragraph under the picture. Be sure your name is on your paper before you hand it in.

Day 5—Review and Reflect

Materials

- "Review and Reflect" form (See below.)

Activity

Review and discuss concepts introduced during the week. Help students complete the form.

Review and Reflect

Name ______________________________ Date ______________

What did all of the dinosaurs you studied this week have in common? ______________

____

____

Illustrations for Week 28

Allosaurus

Tyrannosaurus

Compsognathus

Pteranodon

Background

Continue to put together individual dinosaur books over the course of this unit.

Day 1—Dinosaurs Became Extinct

Materials

- encyclopedias and other current illustrated reference materials
- access to school library, if available

Activity

Have a large group discussion about dinosaurs.

Ask students:

Did people ever see dinosaurs “in person”? (No, dinosaurs became extinct about 60,000,000 years before human beings appeared.)

What does it mean to become extinct? (Get students’ opinions and tell them you will talk about what scientists think about this during the rest of the week.)

Day 2—Doing Some Research

Materials

- classroom wall map of the world
- encyclopedias and other current illustrated reference materials
- access to school library, if available
- research form (see page 89)

Activity

Have students research the different theories on why dinosaurs became extinct, using the form on page 89. Depending on the age group you are teaching, students can do this research themselves or you can lead them through it, reading aloud and having them dictate information to you or to an aide.

Note: One theory holds that the Earth’s terrain changed, setting off a chain of events: mountains rose and the inland seas drained, causing plants and then plant eaters and finally meat eaters to die. Another theory is that a great meteor collided with the Earth, bringing changes in climate.

A still more recent theory is that dinosaurs did not become extinct; they evolved into modern-day birds.

Day 2—Doing Some Research *(cont.)*

Research

Name ______________________________ Date ______________

I found the following theories about why the dinosaurs became extinct:

The first theory is ____

____

____

____.

The second theory is ____

____

____

____.

The third theory is ____

____

____

____.

- -

Day 3—Giving a Report

Materials

- classroom wall map of the world
- encyclopedias and other current illustrated reference materials
- access to school library, if available
- research form (from Day 2)

Activity

Have each student report orally on the information he or she gathered about the dinosaurs becoming extinct. List the various theories on the board. Did everyone find the same information? Did anyone find anything different? Discuss.

Day 4—Dinosaur Fact and Fantasy

Materials

- dinosaur video

Activity

View the video as a group and then discuss it. Which parts were based on fact? Which parts were fantasy?

Day 5—Review and Reflect

Materials

- "Review and Reflect" form (See below.)

Activity

Review and discuss concepts introduced during this week. Help students complete the form. Depending on the age group you are teaching, you may want students to dictate their answers to you or to an aide.

Review and Reflect

Name ______________________________ Date ______________

What I Learned:

What do people mean when they say the dinosaurs became extinct? ______________

____

____

____

What do you think happened to the dinosaurs?______________________

____

____

____

____

Background

Take this opportunity to review and reinforce the concepts and vocabulary introduced during the preceding four weeks.

Day 1—Paleontology

Materials

- encyclopedias and other illustrated reference materials
- *Digging Up Dinosaurs* by Aliki
  (*Let's Read & Find-Out Science Book,* HarperCollins Children's Books, 1988)

Activity

Say to students:

Who remembers what paleontology is? (the science that deals with the study of fossils)

What are scientists who study them called? (paleontologists)

What things do paleontologists do that sound like fun? (camp in wild places, dig in the dirt to look for fossils, look for clues that will lead them to the bones of dinosaurs, and then fit the bones together like giant jigsaw puzzles)

What are fossils and where are they found? (remains of plants and animals that lived in the past—they are found in layers of sedimentary rock)

What are the four main kinds of fossils? (petrified fossils, natural molds, prints, whole animals and plants)

Day 2—Dinosaurs

Materials

- encyclopedias and other illustrated reference materials
- dinosaur projects from Week 26–Week 29

Activity

Take time to finish and display dinosaurs made from projects begun in Week 26. If more time is needed, finish them tomorrow.

Discuss the dinosaurs studied in Week 27.

What do they all have in common? (They are all plant eaters.)

Name them. (Apatosaurus, Stegosaurus, Ankylosaurus, Triceratops)

Day 3—More Dinosaurs

Materials

- encyclopedias and other illustrated reference materials

Activity

If necessary, take time to finish and display dinosaurs made from projects begun in Week 26.

Discuss the dinosaurs studied in Week 28.

What do they all have in common? (They are all meat eaters.)

Name them. (Allosaurus, Tyrannosaurus, Compsognathus, Pteranodon)

Day 4—Still More Dinosaurs

Materials

- construction paper and drawing paper
- crayons, markers, scissors, etc.
- illustrated dinosaur paragraphs from Weeks 27 and 28

Activity

Have students make covers for their dinosaur books and insert the pages made during Week 27 and Week 28. This book can be taken home or included in the students' portfolios.

Say to students:

Who remembers what the word "extinct" means? (no more left)

Did people and dinosaurs ever live at the same time? (No, dinosaurs became extinct about 60,000,000 years before humans appeared.)

Do scientists agree on why dinosaurs became extinct? (No. Discuss theories.)

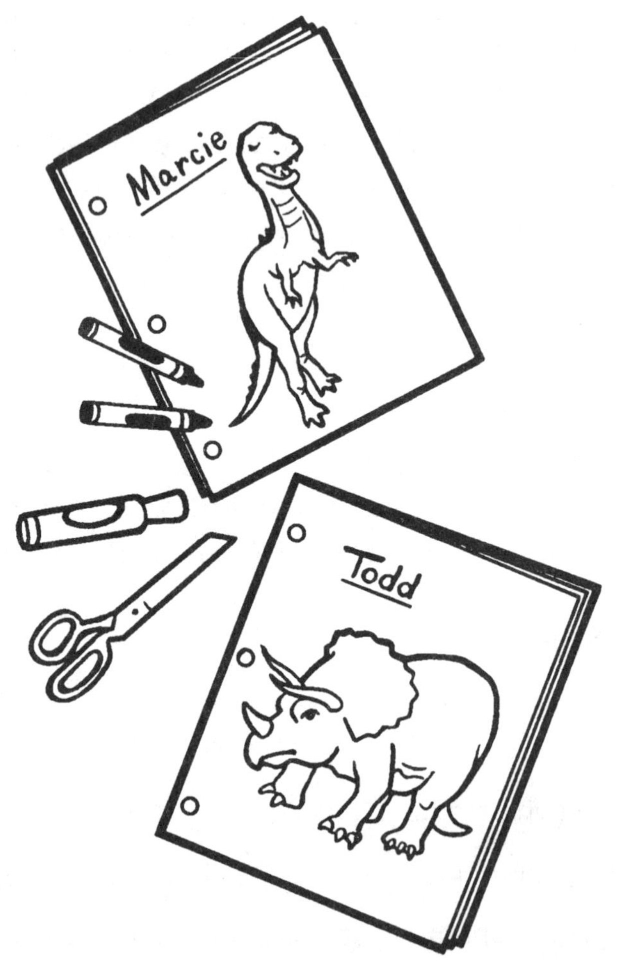

Day 5—Review and Reflect

Materials

- "Review and Reflect" form (See below.)

Activity

Give students necessary help to complete the form below.

Review and Reflect

Name ______________________________ Date ________________

Choose words from below to complete this story.

Paleontologists study ______________________________.

Sometimes the fossils they find are the bones of ______________________________.

Paleontologists dig up the bones and take them back to their ______________________,

where they put them together.

Some dinosaurs were plant eaters. These are the plant eaters we studied:

______________________ ______________________

______________________ ______________________

Other dinosaurs were meat-eaters. These are the meat-eaters we studied:

______________________ ______________________

______________________ ______________________

Dinosaurs have been extinct for ______________________________ of years.

fossils	Tyrannosaurus	museums
Apatosaurus	dinosaurs	Pteranodon
Stegosaurus	millions	Compsognathus
Ankylosaurus	Triceratops	Allosaurus

Background

Learn and discuss some interesting facts about trees.

Day 1—Roots, Trunks, and Branches

Materials

- trees on the playground or at a nearby park
- red and green food coloring
- a stalk of celery with a leafy top
- two tall glasses
- water
- a white carnation

Activity

Take students for a nature walk to observe trees. Have them really look at the trees they see so they will be able to talk about them when they get back to the classroom.

Have them feel the bark on the trunk and pick up a leaf or two.

Pick a small branch with a few leaves attached to talk about later.

Are any of the roots showing above the ground? Have students walk from the trunk out to the edge of the tree where the umbrella of leaves ends. That's about how far the roots extend under the ground.

Back in the classroom, make a simple sketch of a tree trunk, main branches, and roots on the chalkboard. Let student volunteers label them. Where is the bark? Draw an arrow to it and have someone label it, too.

Why is the trunk so thick? (It holds up the branches.) What does the bark do? (It protects the inside of the trunk and branches.) What do the roots do? (They let the tree stand up by grabbing and holding the soil. They also drink water and minerals out of the ground and send them up into the rest of the tree.)

Try this experiment:

- Fill two tall glasses with water.
- Add red food coloring to the water in one glass and green food coloring to the water in the other.
- Cut a thin slice off the bottom of a stalk of celery and put the stalk in the glass with red food coloring.
- Cut a small piece from the bottom of the stem of a carnation and put it in the glass with green food coloring.
- Put the glasses in the sunlight. Leave them overnight.
- Check them in the morning. What do you see? What does it have to do with what we learned about trees? Discuss.

Day 2—Different Leaves

Materials

- broad-leaved and needle-shaped-leaf trees (Find examples on the school grounds or at a nearby park and bring in a small branch from each kind of tree to examine in the classroom.)

Activity

Have students look at, touch, and generally examine both kinds of leaves.

Say to students:

What happens to most broad leaves during the year? (They change color, fall off, and then leaf buds appear and grow again in the spring.)

What happens to needle-shaped-leaf tree? (They stay green all year.)

Who knows another name for needle-shaped-leaf trees? (evergreens)

What job do the leaves do for the tree? (The leaves are the food factories for the trees. Trees take water and minerals from the roots and carbon dioxide from the air. They use their special green cells [chlorophyll] and sunlight to make sugar to feed the tree. Making this sugar causes oxygen to go back into the air. This food-making process is called photosynthesis, which means to make something with light.)

Day 3—Gifts Trees Give Us

Materials

- encyclopedias and other illustrated reference materials
- brainstorming form (See below.)

Activity

Have students meet in groups to brainstorm the things we get from trees. Have them use the form below.

Meet back in a large group to compare and combine lists. (Some examples are fruits, nuts, cocoa, and maple syrup which are used for food; wood which is used to build houses and furniture and burned for heat and fuel; wood pulp which is made into paper; special products such as medicines, cork, rubber, oxygen, etc.)

Gifts Trees Give Us

List everything you can think of that we get from trees.

____

____

Day 4—How People Feel About Trees

Materials

- a book of nature poems

Activity

Read aloud some poems about trees. Some good ones to be found in most anthologies of nature poems are "Year In, Year Out" by Kathleen Millay, "What Do We Plant" by Henry Abbey, and "Trees" by Henry van Dyke. (Probably the most famous tree poem is "Trees" by Joyce Kilmer.)

Have students write poems of their own about trees in general or about one special tree. These poems can be copied on good paper, illustrated, and displayed on a bulletin board for everyone to enjoy.

Day 5—Review and Reflect

Materials

- "Review and Reflect" form (See below.)

Activity

Review and discuss concepts introduced during this week. Help students complete the form.

Review and Reflect

Name ______________________________ Date ______________

What I Learned:

The thick ______________________________ of a tree holds up the branches.

The ______________________________ keep the tree anchored in the soil.

The ______________________________ protects the inside of the tree.

The leaves make ______________ for the tree and ______________ for us.

Trees give us many ______________. I think the most important one is ______________

____.

Background

Use investigation and observation to learn and discuss some interesting facts about seeds.

Day 1—Beans Are Seeds

Materials

- a large bag of dried lima beans

Activity

Tell students that beans are really the seeds of the bean plant.

Pass some dry beans around so everyone can hold, feel, and look at them.

Put at least two beans for each student (plus an extra handful or so) in a bowl, add water to cover, and leave them to soak overnight.

Tell students that they will be finding out more about the beans the next day.

Ask students to think of other seeds that we use for food. (all different kinds of beans, wheat, corn, peas, and so on)

If people eat all these seeds, how do farmers plant more of them to grow? (They save some of the best ones to plant for the next crop.)

Day 2—The Little Plant Inside the Seed

Materials

- bowl of soaked bean seeds from Day 1
- magnifying glasses, at least one for each small group
- seed picture from page 98, enlarged for each student
- paper cups for each student
- soil or planting mix
- marking pens
- plastic trays or cookie sheets lined with soil

Activity

Students can work in small groups to compare and discuss their observations.

Give each student a soaked bean. (Reserve the extras in case the first bean breaks.)

Give students a copy of the seed picture on page 98.

Have students slip off the seed coat, open the seed carefully, and look for the tiny plant on one side of the seed.

Ask students to use the magnifying glass to find little roots, stem, and leaves.

Give each student another soaked bean and a paper cup filled with soil or planting mix.

Have students plant their bean seeds and poke a couple of holes in the bottom of the paper cup with a pencil point. They can write their names on the sides of the cups.

Stand cups on the trays and water. Assign someone to water them each day or let each student water his or her own.

Day 2—The Little Plant Inside the Seed *(cont.)*

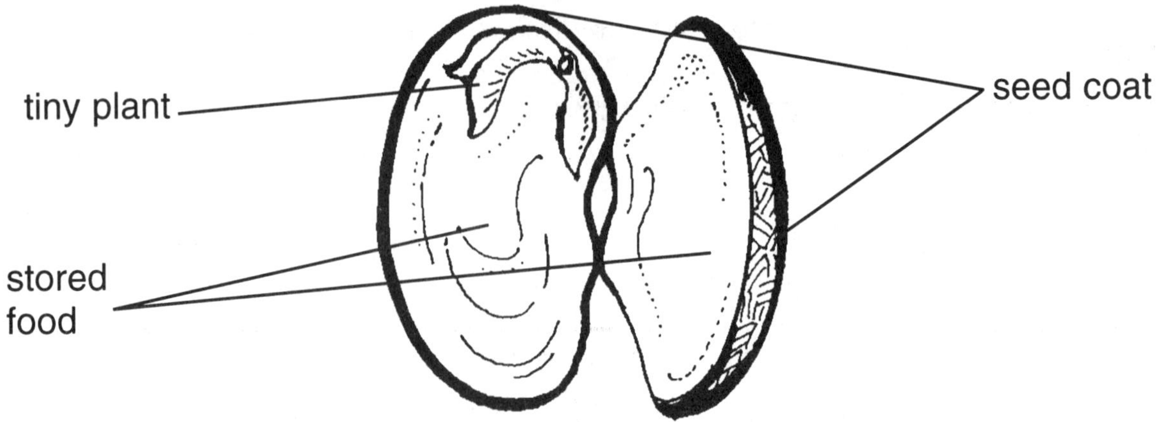

Day 3—How Seeds Travel

Materials

- encyclopedias and other illustrated reference materials
- "How Seeds Travel" sheet enlarged for each student (See below.)

Activity

Depending on the age group you teach, you may wish to help students find this information on how seeds travel or have them do research on their own.

How Seeds Travel

Find an example of each kind of seed and draw and label it below.

Seeds with Wings	Seeds with Hooks	Seeds that Float

Day 4—How Seeds Grow

Materials

- bean seeds plants in paper cups from Day 2
- "My Bean Seed" chart for each student

Activity

Keep a record of what is happening to the bean seeds you planted. Make a chart like this:

My Bean Seed

I planted my bean seed on ______________________________.

On ______________________________, it looked like this:

On ______________________________, it looked like this:

On ______________________________, it looked like this:

Day 5—Review and Reflect

Materials

- "Review and Reflect" form (See below.)

Activity

Review and discuss concepts introduced during this week. Help students complete the form. Depending on the age group you are teaching, you may want students to dictate their answers to you or to an aide.

Review and Reflect

Name ____ Date __________________

The most interesting thing I learned about seeds is ________________________________

____

____

____

Background

Learn and discuss some interesting facts about flowers.

Day 1—The Importance of Flowers

Materials

- encyclopedias and other illustrated reference materials

Activity

Look at pictures of flowers and discuss them.

Say to students:

Human beings depend entirely on flowers and flowering plants for their food supply. Almost all grains, fruits, and vegetables are flowering plants. Even the animals that we use for food live on flowering plants.

Flowers make the seeds for plants. Plants could not make seeds without them.

There are about 200,000 different kinds of flowers. They grow everywhere in the world except in the ice-covered Arctic and Antarctic and in the deep oceans.

Flowers themselves are used as food. We eat flowers when we eat broccoli, cauliflower, asparagus, and artichokes.

Flowers are used as decorations and gifts. Special ones are used on holidays and special occasions. Flowers are also used in industry. Perfume is made from flowers, and so are some dyes.

Day 2—Flowers Make Seeds

Materials

- encyclopedias and other illustrated reference materials
- *Flowers, Fruits, & Seeds* by Jerome Wexler (Simon & Schuster Children's, 1991)

Activity

Look at pictures of flowers and discuss them.

Say to students:

A flower's job is to make seeds. Some flowers make fruit to hold the seeds inside. Other flowers just make seeds inside themselves.

There are many different kinds of flowers. Not all of them have the same parts. Some don't even have petals. Even flowers that are similar have parts that look different.

Pollen helps the flowers to grow seeds. Bees spread pollen from flower to flower as they gather the nectar they use to make honey.

Think of some flowers that don't make fruit. (daisies, pansies, petunias, and so on)

Read *Flowers, Fruits, & Seeds* to the class.

Day 3—Unusual Flowers

Materials

- encyclopedias and other illustrated reference materials
- "My Unusual Flower" (See below.)

Activity

Say to students:

There are many unusual flowers in the world. Look through the reference books we have in the classroom (or in the library) and find the flower you think is the most interesting. Write down its name, where it grows, and something unusual you found out about it. Be ready to share your information with the rest of the class.

My Unusual Flower

My flower's name is ______________________________.

It grows in ______________________________.

I think it is unusual because ______________________________

______________________________.

Day 4—Yarn Flowers

Materials

- encyclopedias and other illustrated reference materials
- paper and pencil
- yarn in different colors
- glue
- scissors
- yarn flower directions and illustrations below and on (page 102)

Activity

Say to students:

Today we are going to make decorative flowers with yarn.

On your paper, use your pencil to draw a light outline of the flower or flowers you would like to make.

Fill in the flowers with pieces of yarn, winding them around until they fill in your flowers.

Glue down the yarn a little at a time so you can make changes if you want as you go along.

Allow the pictures to dry and then display them on a bulletin board.

Day 4—Yarn Flowers *(cont.)*

Day 5—Review and Reflect

Materials

- "Review and Reflect" form (See below.)

Activity

Review and discuss concepts introduced during this week. Help students complete the form.

Review and Reflect

Name ______________________________ Date ______________

What I Learned:

Human beings and animals all depend on flowering plants for ______________________

____

____.

Flowers grow almost ______________________________ in the world.

The main job of flowers is to make ______________________________.

Background

Learn and discuss some interesting facts about fruit.

Day 1—Fruits Are Seed Containers

Materials

- encyclopedias and other illustrated reference materials
- a selection of fruits with different kinds of seeds (or pictures)

Activity

Say to students:

We have looked at seeds, both inside and outside, and planted seeds. We have talked about flowers and how it is their job to make the seeds. Today we are going to look at fruit, the part of many plants that holds the seeds the flowers have made.

Some fruits have lots of seeds all through them. (watermelons)

Some fruits have one big seed in the middle. (peaches)

Some fruits are covered with tiny seeds. (strawberries)

Some fruits have a core containing several seeds. (apples)

Some fruits are all seed. (nuts)

Day 2—Fruits Have a Long History

Materials

- encyclopedias and other illustrated reference materials
- classroom wall map of the world
- peaches (or pictures of peaches)

Activity

Say to students:

People have grown fruits for thousands of years. The peach is a good example of how a fruit has traveled around the world through history.

Have student volunteers point out places on the map as you tell the story of the peach.

Peaches were first grown in China.

Ancient caravans carried peach trees to Persia (now Iran).

Peach trees arrived in Europe when Alexander the Great took them back from Persia to Greece.

Spanish ships carried peach trees from Spain to the New World (America).

Spanish padres planted peach trees around missions in the American Southwest.

Farmers grow peaches that we can enjoy today.

Day 3—Fruit Finger Puppets

Materials

- fruit finger puppets (See below.)
- encyclopedias and other illustrated reference materials

Activity

Have students color and cut out finger puppets of various fruits that hold seeds in different ways.

Encourage students to have their finger puppets tell about themselves: "I am a watermelon. I have lots of seeds all through me." "I am a peach. I have just one seed in my middle. Sometimes it is called a pit or a stone," and so on.

Have a "Fruit Finger Puppet Show."

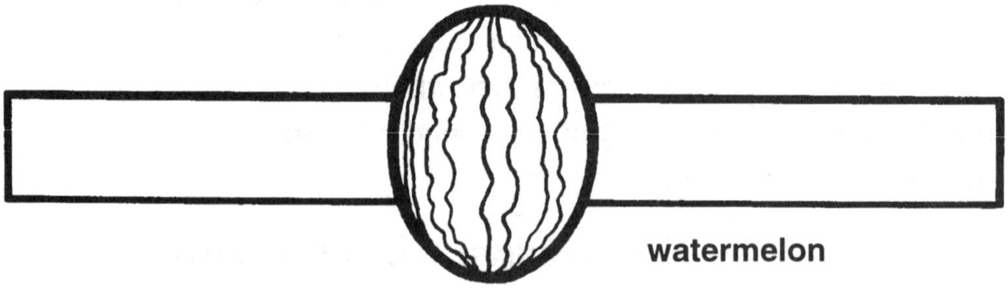

watermelon

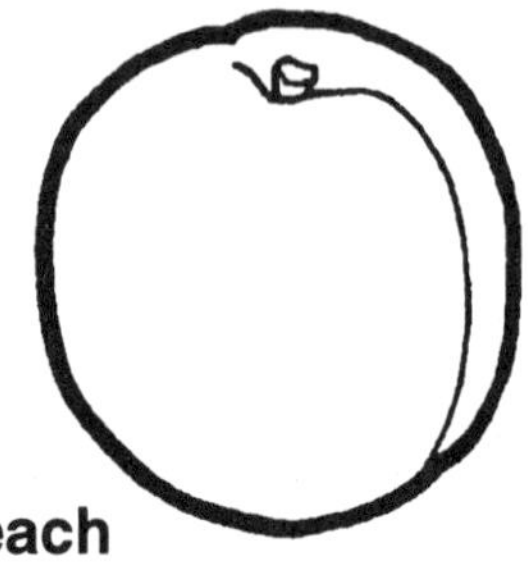

peach

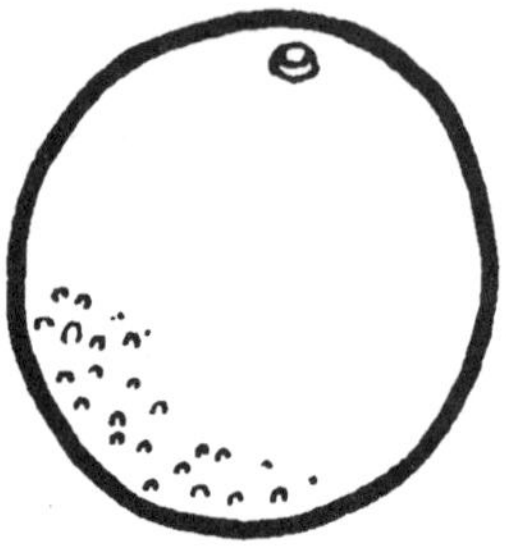

orange

strawberry

apple

grapes

banana

Day 4—Fruit Prints

Materials

- large sheets of white paper
- tempera in fruit colors
- flat pans for dipping fruit
- variety of fruits (lemons, oranges, apples, pears), cut in half

Activity

Put tempera paint into flat pans and give each cooperative group a selection of colors and fruit.

Show students how to dip their fruit in paint and print on the paper. Interesting effects can be obtained by printing first with one color and then with another.

When the prints are dry, press them overnight under a stack of heavy books to flatten them.

Day 5—Review and Reflect

Materials

- "Review and Reflect" form (See below.)

Activity

Review and discuss concepts introduced during this week. Help students complete the form.

Review and Reflect

Name ______________________________ Date ______________

What I Learned:

Fruits are containers for ______________________________.

Watermelons have ______________________________.

Peaches have ______________________________.

Strawberries have ______________________________.

Apples have ______________________________.

Nuts are ______________________________.

People have grown fruits for ______________________________.

Background

Take this opportunity to review and reinforce the concepts and vocabulary introduced during the preceding four weeks.

Day 1—Let's Learn About Trees

Materials

- simple diagram of a tree drawn on the chalkboard

Activity

Ask students:

Who can label the trunk on this tree? Where is the bark? Who can label a branch? Where are the roots?

Why is the trunk so thick? (It holds up the branches.) What does the bark do? (It protects the inside of the trunk and branches.) What do the roots do? (They let the tree stand up by grabbing into and holding the soil. They also drink water and minerals out of the ground and send it up into the rest of the tree.)

Who can describe the experiment we did to show how the roots drink water and send it up to the rest of the tree?

There are two main kinds of leaves. Who can name them?

What job do the leaves do for the tree? (The leaves are the food factories for the trees. They take water and minerals from the roots and carbon dioxide from the air to make sugar for the tree in their green cells [chlorophyll]. Making sugar for the tree causes the leaves to send oxygen into the air, which animals need to live. This process of food-making is called photosynthesis, making something with light.)

Day 2—Let's Learn About Seeds

Materials

- any bean plants that are still in the classroom
- "My Bean Seed" charts

Activity

Ask students:

Who can describe our investigation of bean seeds? Who can add something?

What did we find inside the seeds that were soaked overnight?

What happened to the bean seeds that were planted? Look at your chart and tell the class how your bean plant grew.

Day 3—Let's Learn About Flowers

Materials

- none

Activity

Ask students:

What is a flower's main job? (to make seeds)

Why are flowers so important? (Human beings depend entirely on flowers and flowering plants for their food supply. Almost all grains, fruits, and vegetables are flowering plants. Even the animals that we use for food live on flowering plants.)

How many different kinds of flowers are there? (about 200,000 different kinds)

Where do they grow? (everywhere in the world except in the ice-covered Arctic and Antarctic and in the deep oceans)

What are some things flowers are used for? (food, decorations and gifts, and in industry)

Day 4—Let's Learn About Fruit

Materials

- fruit finger puppets
- classroom wall map of the world

Activity

Ask students:

What is a fruit's main job? (to hold seeds)

There are many different ways that a fruit can get this job done. Use your finger puppets to tell us about how this job is done by the following:

- a watermelon
- a strawberry
- a peach
- an apple

Name another fruit that has. . .

. . . seeds all through it.

. . . one large seed or pit.

. . . nothing but seed.

. . . small seeds on the outside.

. . . a core with several seeds.

Who remembers where peaches first grew?

Who can show us on the map how the peach reached the American Southwest?

Day 5—Review and Reflect

Materials

- "Review and Reflect" form (See below.)

Activity

Give students necessary help to complete the form below.

Review and Reflect

Name ______________________________ Date ______________

Choose words from below to complete this story.

Trees have several important parts.

The __________________ anchor the tree in the ground and drink water for the tree.

The ______________ holds up the branches and sends water up to the rest of the tree.

It is protected by the rough ________________________. The green

___________________________make food for the tree and______________

for us.

Trees grow from ____.

Every seed has a little ___________________________________ inside of it.

Seeds are made by ____.

Many flowers also make some kind of______________________to hold the seeds.

trunk	seeds	flowers
roots	fruit	plant
leaves	bark	oxygen

Background

Learn and discuss some interesting facts about insects.

Day 1—What Do Insects Look Like?

Materials

- encyclopedias and other illustrated reference materials
- enlarged copy of the insect parts below, one for each student

Activity

Say to students:

There are between 2,000,000 and 4,000,000 different kinds of insects in the world.

Some are so tiny they can crawl through the eye of a needle. At least one, the walking stick that lives in the tropics, is a foot long.

They come in all colors of the rainbow. They move in many different ways. But, no matter how different they are, all adult insects have these same basic parts:

1. the head with a brain, feelers (antennae), eyes, and mouth
2. the middle body or thorax with 6 legs and 2 sets of wings
3. the abdomen where food is digested

Cut out the parts below. Put them together by gluing them in the right places on a piece of construction paper.

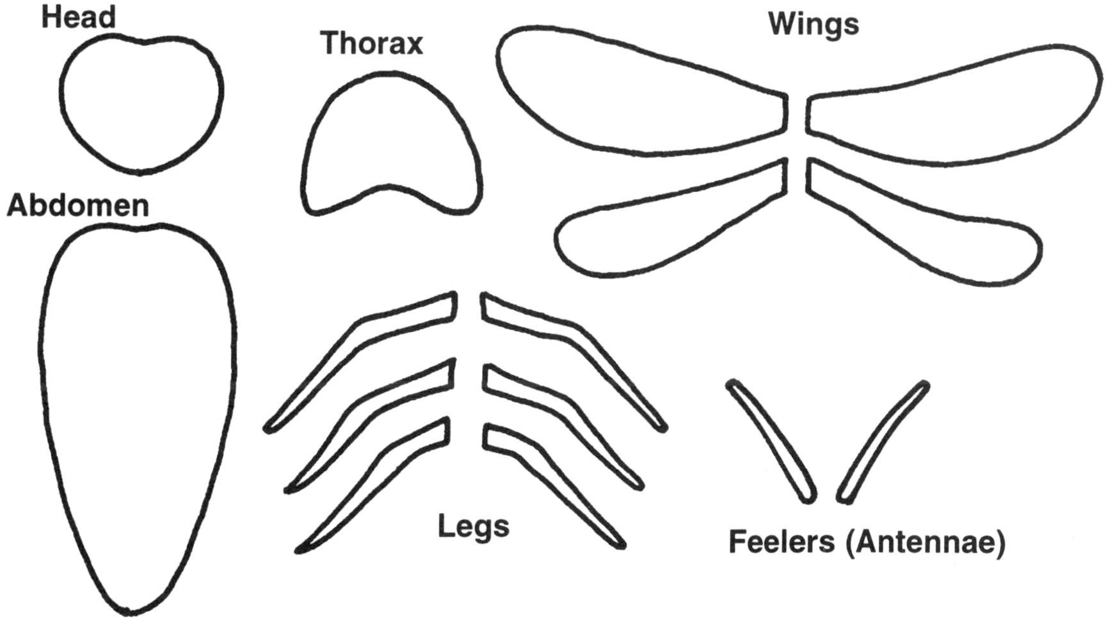

Day 2—Researching Insects from A to Z

Materials

- encyclopedias and other illustrated reference materials
- a trip to the school library, if possible
- "Insect Alphabet" form (See below.)

Activity

Assign each student a different letter of the alphabet.

Challenge them to find an insect whose name begins with the given letter and then complete the form below. (Depending on the age group you are teaching, you may want to help students with this research or have them do it on their own.)

Insect Alphabet

Name ______________________________ Date ______________

My letter of the alphabet is ______________________________.

My insect is called ______________________________.

Its size is ______________________________.

It lives ______________________________.

This is what I learned about it:

____

____

____

Day 3—Reports on Insects from A to Z

Materials

- completed "Insect Alphabet" form from Day 2

Activity

Have each student give his or her report in alphabetical order. Put the forms together in an "Insects from A to Z" book and add it to your classroom library.

Day 4—Making an Insect Model

Materials

- encyclopedias and other illustrated reference materials
- finished insects from Day 1
- cardboard egg cartons
- heavy paper or cardboard
- construction paper
- pipe cleaners
- paint or colored markers

Activity

Say to students:

Make your own insect, using sections from an egg carton for the head, thorax, and abdomen. You may make legs from heavy paper, wings from construction paper, and feelers from pipe cleaners.

Use your insect-parts picture from Day 1 to help you remember where the parts go.

Use your imagination to make your insect different from all the others.

Day 5—Review and Reflect

Materials

- "Review and Reflect" form (See below.)

Activity

Review and discuss concepts introduced during this week. Help students complete the form. Depending on the age group you are teaching, you may want students to dictate their answers to you or to an aide.

Review and Reflect

Name ________________________________ Date ______________

What I Learned:

How many different kinds of insects are there in the world? ______________________

How many main body parts do adult insects have? ______________________

How many legs do they have? ______________________

How many sets of wings do they have? ______________________

Background

Learn and discuss some interesting facts about reptiles.

Day 1—Alligators and Crocodiles

Materials

- encyclopedias and other illustrated reference materials
- "Alligators and Crocodiles" form (See below.)

Activity

Say to students:

Reptiles are related to the dinosaurs. The dinosaurs were reptiles, too.

There are four different kinds of reptiles:

- alligators and crocodiles
- snakes
- lizards
- turtles

They all have backbones.

They are all cold blooded, which means they can't control their temperatures. They become the same as the temperature around them.

Most of them hatch their young from eggs.

Have students research the following questions. Depending on the age group you are teaching, you may want to help students with this research or have them do it on their own.

Alligators and Crocodiles

What are some differences between alligators and crocodiles?

____

____

Where do they live? ____

____

____

____

Day 2—Lizards

Materials

- encyclopedias and other illustrated reference materials
- "Unusual Lizards" list (See below.)

Activity

Have students research these unusual lizards. Depending on the age group you are teaching, you may want to help students with this research or have them do it on their own.

Unusual Lizards

Find out something about each of these unusual lizards. Use another piece of paper to write the descriptions on.

- Australian fringed lizard
- Komodo dragon
- glass snake
- chameleon
- flying dragon
- horned toad
- Gila monster

Day 3—Snakes

Materials

- encyclopedias and other illustrated reference materials

Activity

Have students research poisonous snakes in the United States. Ask them to find out how many there are, what their names are, and how they can be identified. Depending on the age group you are teaching, you may want to help students with this research or have them do it on their own.

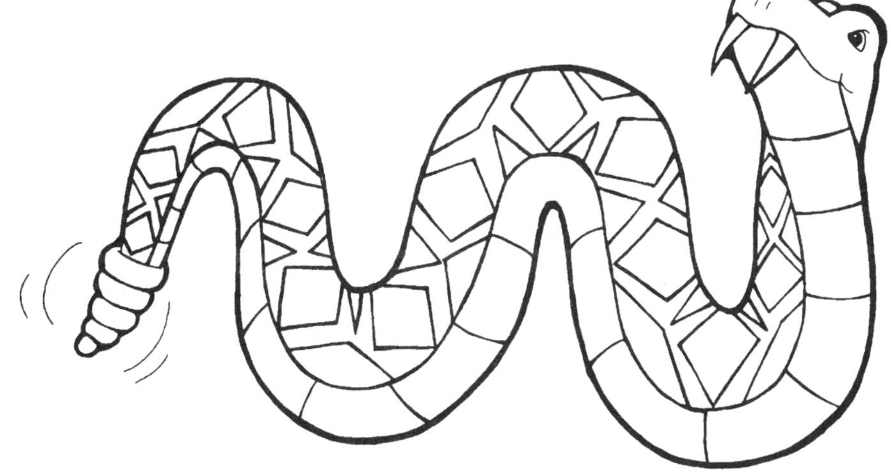

Day 4—Turtles

Materials

- encyclopedias and other illustrated reference materials
- a video about turtles

Activity

Give students the following information. Then show the video and discuss it.

Turtles have been around for the past 175,000,000 years. During that time the dinosaurs and many other creatures came and went.

Some people call desert turtles "tortoises" and sea turtles "terrapins," but they can all correctly be called turtles.

Some turtles live only in the water, some live only on land, and some spend part of their time in both places.

The thing that makes turtles different from other reptiles, and from other animals, is the shell or "house" they carry on their backs.

Unlike other animals with backbones (vertebrates), turtles do not have teeth.

Turtles also live longer than any other animal with a backbone.

Day 5—Review and Reflect

Materials

- "Review and Reflect" form (See below.)

Activity

Review and discuss concepts introduced during this week. Help students complete the form. Depending on the age group you are teaching, you may want students to dictate their answers to you or to an aide.

Review and Reflect

Name ______________________________ Date ______________

What I Learned:

Lizards, snakes, alligators, crocodiles, and turtles are all ______________________.

The ____ were reptiles too.

Reptiles are interesting because ____

____.

Background

Learn and discuss some interesting facts about amphibians.

Day 1—Some Facts About Frogs

Materials

- encyclopedias and other illustrated reference materials

Activity

Say to students:

Frogs are amphibians. Amphibians are animals that spend part of their lives in the water and part of their lives on land.

Frogs come in many sizes. The giant frog of Africa is a foot long and weighs as much as a small dog. On the other hand, some frogs are smaller than a penny!

Frogs have long back legs for jumping and webbed toes for swimming.

Frogs can shoot out their long sticky tongues to capture insects.

Tree frogs have little suckers on their feet that help them climb.

Many frogs have poison in their bodies. The poison oozes out of the skin and helps to protect them from animals that want to eat them.

Day 2—Life Stages of Frogs

Materials

- encyclopedias and other illustrated reference materials
- study prints or an appropriate video

Activity

Look at the pictures or video and discuss.

Say to students:

Frogs lay their eggs in the water. The eggs look like little balls of jelly.

Tiny animals called tadpoles or polliwogs hatch from the eggs. Tadpoles have gills like fish.

After awhile, tadpoles grow legs. The hind legs develop first. The gills disappear, and the tadpoles must come to the surface to breathe air.

The tadpole's long tail is absorbed into its body, and it hops out onto land as an adult frog.

Day 3—Sequence the Stages

Materials

- encyclopedias and other illustrated reference materials
- pictures of stages in the life of the frog (See below.) enlarged for each student
- paper

Activity

Color the pictures below. Then cut them out and glue them on another sheet of paper in the right order.

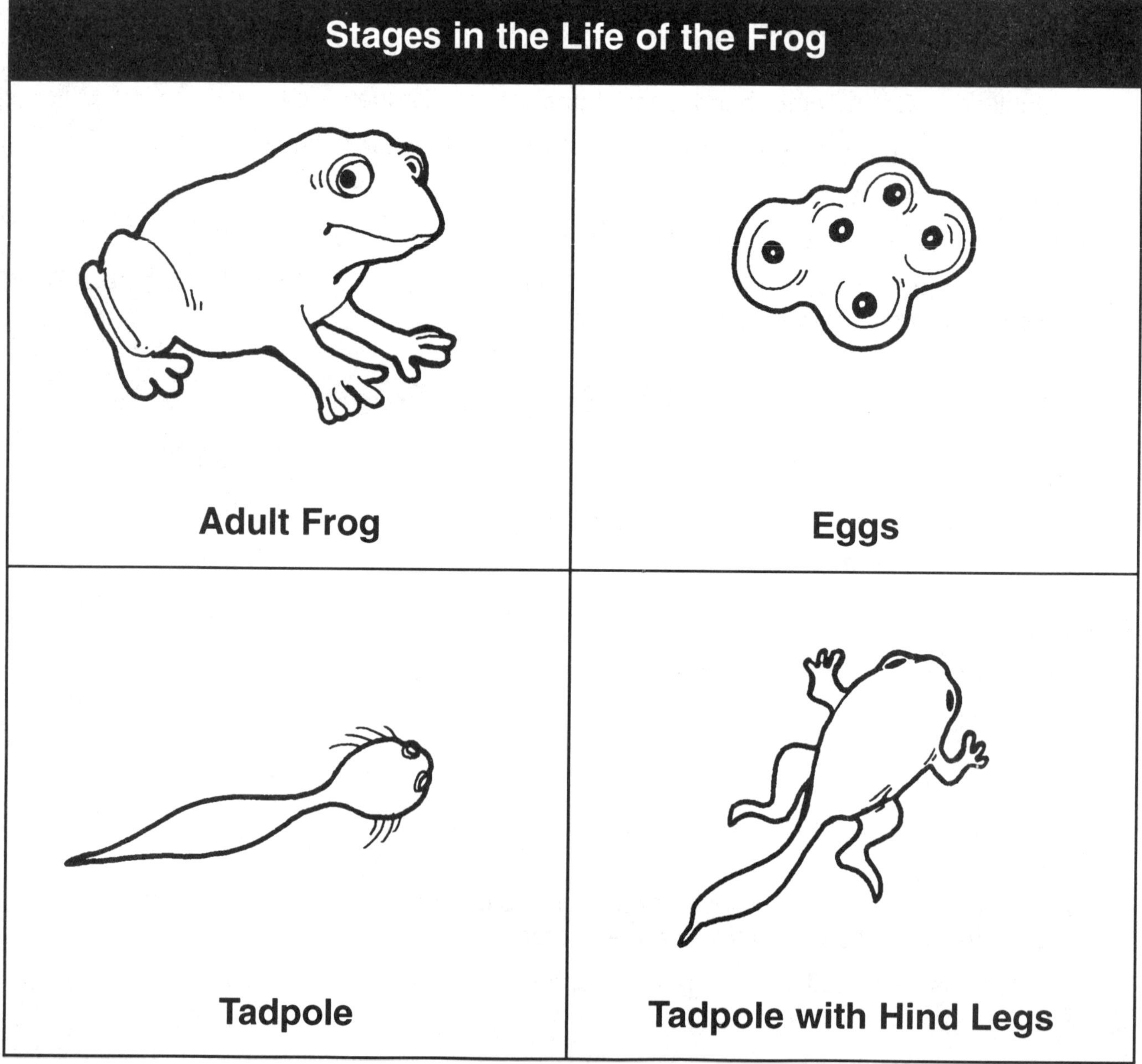

Day 4—Frogs Are Important to People

Materials

- encyclopedias and other illustrated reference materials
- "Frogs Are Important to People" research form (See below.)

Activity

Have students find out ways in which frogs are important to people. Depending on the age group you are teaching, you may want to help students with this research or have them do it on their own.

Frogs Are Important to People

Frogs are important to people in these ways:

____

____

____

Day 5—Review and Reflect

Materials

- "Review and Reflect" form (See below.)

Activity

Review and discuss concepts introduced during this week. Help students complete the form. Depending on the age group you are teaching, you may want students to dictate their answers to you or to an aide.

Review and Reflect

Name ____________________________ Date ______________

What I Learned:

The most interesting thing I learned about frogs was ____________________

____

____.

Background

Learn and discuss some interesting facts about fish.

Day 1—Fish Facts

Materials

- encyclopedias and other illustrated reference materials
- "Fact Fish" (See below.)

Activity

Copy the "Fact Fish" pattern below on colored construction paper. Make one for each student in your class.

On the front, write the name of an unusual or interesting fish. (See suggestions below.)

Have each student look up his or her fish and write one fact about it on the back. Depending on the age group you are teaching, you may want to help students with this research or let them do it on their own.

Collect all the Fact Fish and play the following Fish Facts game:

Read the fact aloud. Call on someone to name the fish. If the answer is correct, that person holds the fish. The student with the most fish at the end of the game wins.

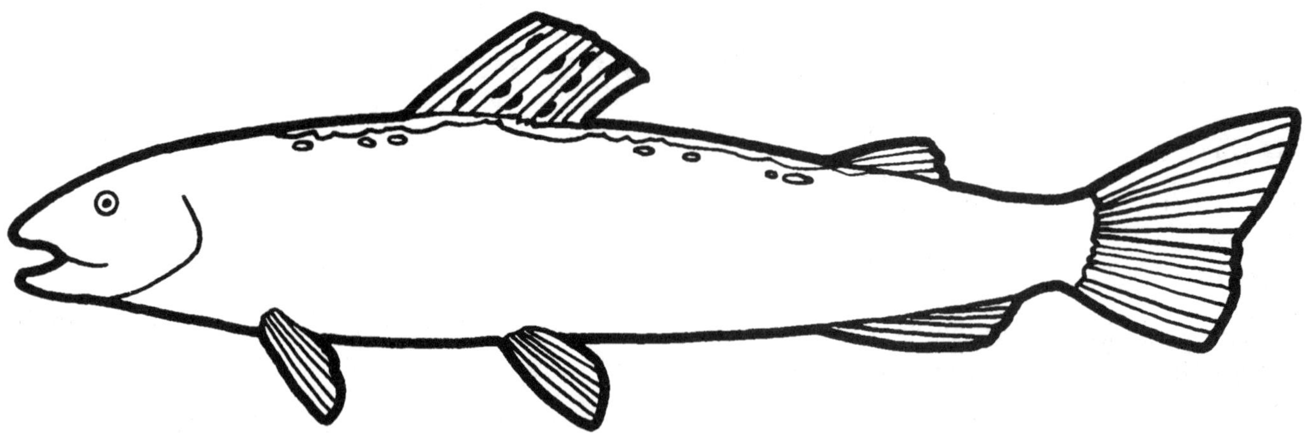

1. archer fish
2. common puffer
3. walking fish
4. scavenger fish
5. flying fish
6. kissing fish
7. black swallower
8. four-eyed fish
9. angler fish
10. dragon eel
11. tree climber
12. glass fish
13. convoy fish
14. pipe fish
15. brook stickleback
16. butterfish
17. banjo catfish
18. Gaff-Topsail catfish
19. Siamese fighting fish
20. Hawaiian butterfly fish
21. lion fish
22. sea anemone fish
23. neon tetra
24. swordtail
25. lyretail
26. sailfish
27. sea horse
28. electric eel
29. piranha
30. sawfish

Day 2—More Fish Facts

Materials

- encyclopedias and other illustrated reference materials
- blank "Fact Fish" (from pattern on Day 1)

Activity

Copy the "Fact Fish" pattern on colored construction paper. Make a stack of blank ones.

Let students make as many Fact Fish as they can, writing the name of the fish on one side and an interesting fact about the fish on the other.

Put the Fact Fish aside along with the first set to use later in the week.

Day 3—Crayon Resist Fish

Materials

- encyclopedias and other illustrated reference materials
- white drawing paper
- crayons
- thin blue tempera or watercolor

Activity

Discuss the fish that students have seen in the reference books they have used. Talk about their colors and shapes. Ask if anyone remembers a particularly beautiful fish.

Have students draw a variety of fish of their choosing on the white drawing paper and color them in heavily with crayon. They can also add seaweed and other plants.

When the drawings are complete, have students cover the whole paper with a thin blue wash to look like water.

Day 4—Play the Fish Facts Game

Materials

- Fact Fish created on Days 1 and 2

Activity

Spend the period playing as many rounds of Fish Facts as you have time for.

Consider giving prizes to the winners. ("Out-to-Lunch-Early" passes make great prizes.)

Day 5—Review and Reflect

Materials

- "Review and Reflect" form (See below.)

Activity

Review and discuss concepts introduced during this week. Help students complete the form. Depending on the age group you are teaching, you may want students to dictate their answers to you or to an aide.

Review and Reflect

Name ______________________________ Date ______________

What I Learned:

Do fish have backbones?	Yes	No
Do fish eat other fish?	Yes	No
Do fish lay eggs?	Yes	No
Do fish live in fresh water?	Yes	No
Do fish live in salt water?	Yes	No
Do fish live in the tropics?	Yes	No
Do fish breathe air?	Yes	No

Week 40

Review Lesson

Background

Take this opportunity to review and reinforce the concepts and vocabulary introduced during the preceding four weeks.

Day 1—Let's Learn About Insects

Materials

- encyclopedias and other illustrated reference materials
- sketch on chalkboard of main parts of insects

Activity

Ask students:

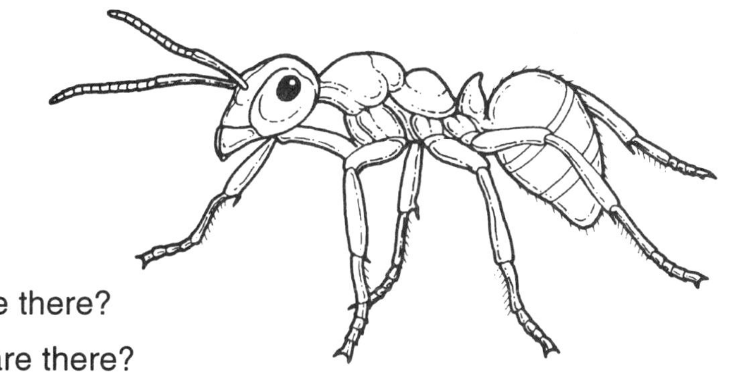

Who can label the head?

Who can label the thorax?

Who can label the abdomen?

Who can label the legs? How many are there?

Who can label the wings? How many are there?

Who can label the feelers? What is another word for feelers?

Day 2—Let's Learn About Reptiles

Materials

- encyclopedias and other illustrated reference materials

Activity

Ask students:

Who can name four different kinds of reptiles?

Who can tell us something about alligators and crocodiles?

Who can tell us something about lizards?

Who can tell us something about snakes?

Who can tell us something about turtles?

Do all reptiles have backbones?

Do most of them hatch their young from eggs?

Who will tell us about an interesting reptile they studied?

Use the time you have left to let as many students as possible share.

Day 3—Let's Learn About Amphibians

Materials

- encyclopedias and other illustrated reference materials

Activity

Ask students:

Which kind of amphibian did we study? (the frog)

What are amphibians? (animals that spend part of their lives in the water and part of their lives on land)

Why do frogs jump so well? (long back legs)

Why do frogs swim so well? (webbed toes for swimming)

How do frogs catch insects? (by shooting out their long, sticky tongues)

What are the four main stages in a frog's life? (egg, tadpole, tadpole with legs, adult frog)

Display and discuss students' sequenced stages of a frog from Week 38, Day 3.

Day 4—Let's Learn About Fish

Materials

- encyclopedias and other illustrated reference materials

Activity

Review with students:

Are fish vertebrates? What is a vertebrate? (yes—animals with backbones)

Do fish eat other fish? (yes)

Do fish lay eggs? (yes)

Do fish live in lakes? (yes)

Do fish live in the ocean? (yes)

Do fish live in warm climates? (yes)

Do fish live in cold climates? (yes)

What do fish use to breathe? (gills)

Play Fish Facts:

Give the name of the fish and ask for a fact.

Give a fact and ask for the name of the fish.

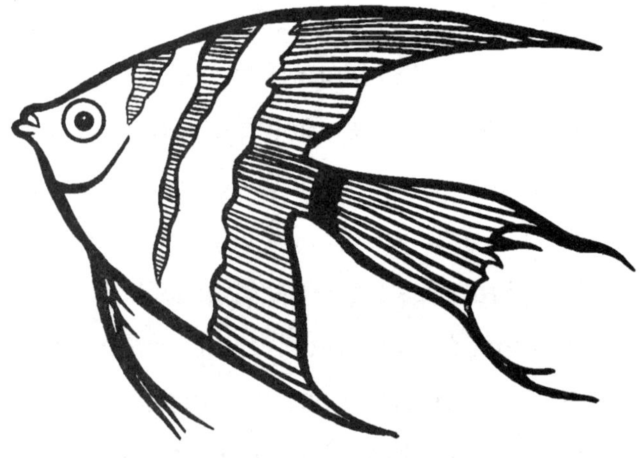

Day 5—Review and Reflect

Materials

- "Review and Reflect" form (See below.)

Activity

Give students necessary help to complete the form below. Depending on the age group you are teaching, they can do it themselves or dictate their information to you or to an aide.

Review and Reflect

Name ______________________________ Date ______________

Choose words from the list below to complete this story.

Insects have ________________ main body parts.

They are the ________________, the ________________,

and the ________________. Reptiles are vertebrates.

They all have a ________________. The reptiles we studied were alligators

or crocodiles, ________________ and ________________.

Amphibians and fish are ________________, too, because they have backbones.

The amphibian we studied was the ________________.

All of these kinds of animals lay ________________.

All of them have at least some members that are ________________.

backbone
eggs
abdomen
three
lizards
snakes

vertebrates
frog
poisonous
thorax
head

Week 41 — Let's Learn About Mammals

Background

Plan to put together individual books over the course of the next few weeks.

Day 1—Lions

Materials

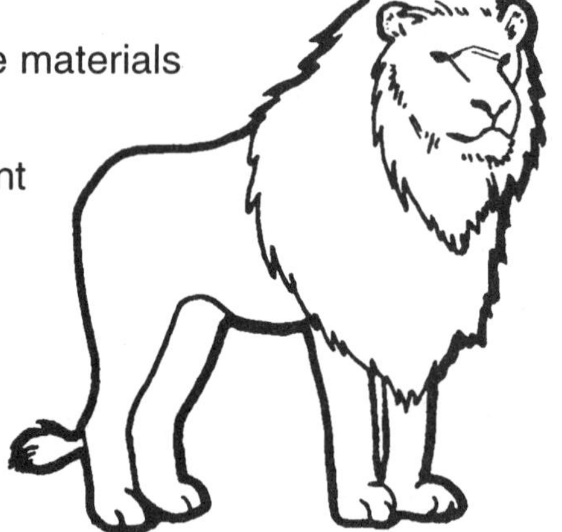

- encyclopedias and other up-to-date illustrated reference materials
- school library access, if possible
- enlarged picture of the lion on page 126 for each student
- writing materials
- drawing and coloring materials

Activity

Say to students:

Color and cut out the picture of the lion.

Use reference books to get information about lions and write a paragraph about them. (Depending on the age group you are teaching, you may want to read the information aloud to students and have them dictate their paragraphs to you.)

Mount your colored picture at the top of a piece of paper. Copy your paragraph under the picture. Be sure your name is on your paper before you hand it in.

Day 2—Bears

Materials

- encyclopedias and other up-to-date illustrated reference materials
- school library, if available
- enlarged picture of the bear on page 126 for each student
- writing materials
- drawing and coloring materials

Activity

Say to students:

Color and cut out the picture of the bear.

Use reference books to get information about bears and write a paragraph about them. (Depending on the age group you are teaching, you may want to read the information aloud to students and have them dictate their paragraphs to you.)

Mount your colored picture at the top of a piece of paper. Copy your paragraph under the picture. Be sure your name is on your paper before you hand it in.

Day 3—Giraffes

Materials

- encyclopedias and other up-to-date illustrated reference materials
- school library access, if possible
- enlarged picture of the giraffe on page 126 for each student
- writing materials
- drawing and coloring materials

Activity

Say to students:

Color and cut out the picture of the giraffe.

Use reference books to get information about giraffes and write a paragraph about them. (Depending on the age group you are teaching, you may want to read the information aloud to students and have them dictate their paragraphs to you.)

Mount your colored picture at the top of a piece of paper. Copy your paragraph under the picture. Be sure your name is on your paper before you hand it in.

Day 4—Elephants

Materials

- encyclopedias and other up-to-date illustrated reference materials
- school library access, if possible
- enlarged picture of the elephant on page 126 for each student
- writing materials
- drawing and coloring materials

Activity

Say to students:

Color and cut out the picture of the elephant.

Use reference books to get information about elephants and write a paragraph about them.

(Depending on the age group you are teaching, you may want to read the information aloud to students and have them dictate their paragraphs to you.)

Mount your colored picture at the top of a piece of paper. Copy your paragraph under the picture. Be sure your name is on your paper before you hand it in.

Day 5—Review and Reflect

Materials

- "Review and Reflect" form (See below.)

Activity

Review and discuss concepts introduced during the week. Help students complete the form. Depending on the age group you are teaching, you may want students to dictate their answers to you or to an aide.

Review and Reflect

Name ______________________________ Date ______________

What do all of the animals you studied this week have in common?

____

Illustrations for Week 41

Lion

Bear

Giraffe

Elephant

Week 42

Let's Learn About More Mammals

Background

Plan to put together individual books over the course of the next few weeks.

Day 1—Cows

Materials

- encyclopedias and other up-to-date illustrated reference materials
- school library access, if possible
- enlarged picture of the cow on page 129 for each student
- writing materials
- drawing and coloring materials

Activity

Say to students:

Color and cut out the picture of the cow.

Use reference books to get information about cows and write a paragraph about them.

(Depending on the age group you are teaching, you may want to read the information aloud to students and have them dictate their paragraphs to you.)

Mount your colored picture at the top of a piece of paper. Copy your paragraph under the picture. Be sure your name is on your paper before you hand it in.

Day 2—Sheep

Materials

- encyclopedias and other up-to-date illustrated reference materials
- school library access, if possible
- enlarged picture of the sheep on page 129 for each student
- writing materials
- drawing and coloring materials

Activity

Say to students:

Color and cut out the picture of the sheep.

Use reference books to get information about sheep and write a paragraph about them.

(Depending on the age group you are teaching, you may want to read the information aloud to students and have them dictate their paragraphs to you.)

Mount your colored picture at the top of a piece of paper. Copy your paragraph under the picture. Be sure your name is on your paper before you hand it in.

Day 3—Pigs

Materials

- encyclopedias and other up-to-date illustrated reference materials
- school library access, if possible
- enlarged picture of the pig on page 129 for each student
- writing materials
- drawing and coloring materials

Activity

Say to students:

Color and cut out the picture of the pig.

Use reference books to get information about pigs and write a paragraph about them.

(Depending on the age group you are teaching, you may want to read the information aloud to students and have them dictate their paragraph to you.)

Mount your colored picture at the top of a piece of paper. Copy your paragraph under the picture. Be sure your name is on your paper before you hand it in.

Day 4—Cats and Dogs

Materials

- encyclopedias and other up-to-date illustrated reference materials
- school library access, if possible
- enlarged picture of the cat and dog on page 129 for each student
- writing materials
- drawing and coloring materials

Activity

Say to students:

Color and cut out the picture of the cat and dog.

Use reference books to get information about cats and dogs and write a paragraph about them.

(Depending on the age group you are teaching, you may want to read the information aloud to students and have them dictate their paragraphs to you.)

Mount your colored picture at the top of a piece of paper. Copy your paragraph under the picture. Be sure your name is on your paper before you hand it in.

Day 5—Review and Reflect

Materials

- "Review and Reflect" form (See below.)

Activity

Review and discuss concepts introduced during the week. Help students complete the form. Depending on the age group you are teaching, you may want students to dictate their answers to you or to an aide.

Review and Reflect

Name ______________________________________ Date ________________

What do all of the animals you studied this week have in common?

____

Illustrations for Week 42

Cow

Sheep

Pig

Cat and Dog

Background

Plan to put together individual books over the course of the next few weeks.

Day 1—Kangaroos

Materials

- encyclopedias and other up-to-date illustrated reference materials
- school library access, if possible
- enlarged picture of the kangaroo on page 132 for each student
- writing materials
- drawing and coloring materials

Activity

Say to students:

Color and cut out the picture of the kangaroo.

Use reference books to get information about kangaroos and write a paragraph about them.

(Depending on the age group you are teaching, you may want to read the information aloud to students and have them dictate their paragraphs to you.)

Mount your colored picture at the top of a piece of paper. Copy your paragraph under the picture. Be sure your name is on your paper before you hand it in.

Day 2—Koalas

Materials

- encyclopedias and other up-to-date illustrated reference materials
- school library access, if possible
- enlarged picture of the koalas on page 132 for each student
- writing materials
- drawing and coloring materials

Activity

Say to students:

Color and cut out the picture of the koala.

Use reference books to get information about koalas and write a paragraph about them.

(Depending on the age group you are teaching, you may want to read the information aloud to students and have them dictate their paragraphs to you.)

Mount your colored picture at the top of a piece of paper. Copy your paragraph under the picture. Be sure your name is on your paper before you hand it in.

Day 3—Tasmanian Devils

Materials

- encyclopedias and other up-to-date illustrated reference materials
- school library access, if possible
- enlarged picture of the Tasmanian devil on page 132 for each student
- writing materials
- drawing and coloring materials

Activity

Say to students:

Color and cut out the picture of the Tasmanian devil.

Use reference books to get information about Tasmanian devils and write a paragraph about them. (Depending on the age group you are teaching, you may want to read the information aloud to students and have them dictate their paragraphs to you.)

Mount your colored picture at the top of a piece of paper. Copy your paragraph under the picture. Be sure your name is on your paper before you hand it in.

Day 4—Opossum

Materials

- encyclopedias and other up-to-date illustrated reference materials
- school library access, if possible
- enlarged picture of the opossum on page 132 for each student
- writing materials
- drawing and coloring materials

Activity

Say to students:

Color and cut out the picture of the opossum.

Use reference books to get information about the opossum and write a paragraph about this animal.

(Depending on the age group you are teaching, you may want to read the information aloud to students and have them dictate their paragraphs to you.)

Mount your colored pictures at the top of a piece of paper. Copy your paragraph under the picture. Be sure your name is on your paper before you hand it in.

Day 5—Review and Reflect

Materials

- "Review and Reflect" form (See below.)

Activity

Review and discuss concepts introduced during the week. Help students complete the form.

Depending on the age group you are teaching, you may want students to dictate their answers to you or to an aide.

Review and Reflect

Name ______________________________ Date ____________

What do all of the animals you studied this week have in common?

____

Illustrations for Week 43

Kangaroo

Koala

Tasmanian Devil

Opossum

Background

Plan to put together individual books over the course of the next few weeks.

Day 1—Robins

Materials

- encyclopedias and other up-to-date illustrated reference materials
- school library access, if possible
- enlarged picture of the robin on page 135 for each student
- writing materials
- drawing and coloring materials

Activity

Say to students:

Color and cut out the picture of the robin.

Use reference books to get information about robins and write a paragraph about them.

(Depending on the age group you are teaching, you may want to read the information aloud to students and have them dictate their paragraphs to you.)

Mount your colored picture at the top of a piece of paper. Copy your paragraph under the picture. Be sure your name is on your paper before you hand it in.

Day 2—Hummingbirds

Materials

- encyclopedias and other up-to-date illustrated reference materials
- school library access, if possible
- enlarged picture of the hummingbird on page 135 for each student
- writing materials
- drawing and coloring materials

Activity

Say to students:

Color and cut out the picture of the hummingbird.

Use reference books to get information about hummingbirds and write a paragraph about them. (Depending on the age group you are teaching, you may want to read the information aloud to students and have them dictate their paragraphs to you.)

Mount your colored picture at the top of a piece of paper. Copy your paragraph under the picture. Be sure your name is on your paper before you hand it in.

Day 3—Woodpeckers

Materials

- encyclopedias and other up-to-date illustrated reference materials
- school library access, if possible
- enlarged picture of the woodpecker on page 135 for each student
- writing materials
- drawing and coloring materials

Activity

Say to students:

Color and cut out the picture of the woodpecker.

Use reference books to get information about woodpeckers and write a paragraph about them. (Depending on the age group you are teaching, you may want to read the information aloud to students and have them dictate their paragraphs to you.)

Mount your colored picture at the top of a piece of paper. Copy your paragraph under the picture. Be sure your name is on your paper before you hand it in.

Day 4—Eagle

Materials

- encyclopedias and other up-to-date illustrated reference materials
- school library access, if possible
- enlarged picture of the eagle on page 135 for each student
- writing materials
- drawing and coloring materials

Activity

Say to students:

Color and cut out the picture of the eagle.

Use reference books to get information about the eagle and write a paragraph about this bird.

(Depending on the age group you are teaching, you may want to read the information aloud to students and have them dictate their paragraphs to you.)

Mount your colored picture at the top of a piece of paper. Copy your paragraph under the picture. Be sure your name is on your paper before you hand it in.

Day 5—Review and Reflect

Materials

- "Review and Reflect" form (See below.)

Activity

Review and discuss concepts introduced during the week. Help students complete the form. Depending on the age group you are teaching, you may want students to dictate their answers to you or to an aide.

Review and Reflect

Name ______________________________ Date ______________

What do all of the animals you studied this week have in common?

____

Illustrations for Week 44

Background

Take this opportunity to review and reinforce the concepts and vocabulary introduced during the preceding four weeks.

Day 1—Let's Learn About Mammals

Materials

- encyclopedias and other illustrated reference materials

Activity

Ask students:

Which animals did we study during the first week about mammals? (lion, bear, giraffe, and elephant)

What did you learn about lions? (Answers will vary.)

What did you learn about bears? (Answers will vary.)

What did you learn about giraffes? (Answers will vary.)

What did you learn about elephants? (Answers will vary.)

Which animal do you think is the most interesting?

Write the names of the animals on the chalkboard and have each student stand in front of the name of his or her most interesting animal for a living line graph.

What do all these animals have in common? (They are all wild animals.)

Day 2—Let's Learn About More Mammals

Materials

- encyclopedias and other illustrated reference materials

Activity

Ask students:

Which animals did we study during the second week about mammals? (cow, sheep, pig, and cat and dog)

What did you learn about cows? (Answers will vary.)

What did you learn about sheep? (Answers will vary.)

What did you learn about pigs? (Answers will vary.)

What did you learn about cats and dogs? (Answers will vary.)

Which animal do you think is the most interesting?

Write the names of the animals on the chalkboard and have each student stand in front of the name of his or her most interesting animal for a living line graph.

What do all these animals have in common? (They are all domesticated animals.)

Day 3—Let's Learn About Marsupials

Materials

- encyclopedias and other illustrated reference materials

Activity

Ask students:

Which marsupials did we study? (kangaroo, koala, Tasmanian devil, opossum)

What did you learn about kangaroos? (Answers will vary.)

What did you learn about koalas? (Answers will vary.)

What did you learn about Tasmanian devils? (Answers will vary.)

What did you learn about opossums? (Answers will vary.)

What makes an animal a marsupial? (It keeps its young in a pouch.)

Where do most marsupials live? (Australia and New Zealand)

Day 4—Let's Learn About Birds

Materials

- encyclopedias and other illustrated reference materials
- construction paper/drawing paper
- crayons, markers, scissors, etc.
- illustrated paragraphs from Weeks 41–44

Activity

Have students make covers for their animal books and insert the pages made during Weeks 41–44. This book can be taken home or included in the student's portfolios.

Ask students:

Which birds did we study? (robin, hummingbird, woodpecker, eagle)

What did you learn about robins? (Answers will vary.)

What did you learn about hummingbirds? (Answers will vary.)

What did you learn about woodpeckers? (Answers will vary.)

What did you learn about eagles? (Answers will vary.)

What can birds do that other animals can't? (Most of them can fly.)

Day 5—Review and Reflect

Materials

- "Review and Reflect" form (See below.)

Activity

Give students necessary help to complete the form below. Depending on the age group you are teaching, they can do it themselves or dictate their information to you or to an aide.

Review and Reflect

Name ______________________________ Date ______________

Choose words from below to complete this story.

Lions, bears, giraffes, and elephants are all ______________________ animals.

We usually see them only in ______________________________.

______________, ______________ and ______________ usually live on farms.

______________________________ carry their young in pouches.

Only the ______________________________ lives in North America.

Most birds can ______________________________.

They have ______________________ and lay ______________________.

zoos	feathers	eggs
fly	sheep	wild
cows	pigs	opossum
		marsupials

The most interesting thing I learned about animals is ______________________

____

____

____

____

____

Background

Young children tend to get upset about threats to the environment. Try to keep the focus on things that they can actually do to help without dwelling on scary pollution statistics and dire predictions.

Day 1—Clean Beaches

Materials

- trash bags
- poster paint
- poster board or large sheets of paper
- brushes
- marking pens

Activity

Say to students:

People worry a lot about keeping our beaches and shores clean and beautiful. You may not have to worry if you decide to do something about it.

Here are some things to do:

Always pick up every bit of your own trash. I have some trash bags available so you can be ready.

Always pick up a little bit extra because some people forget. The waves also throw trash on the beach.

Ask the lifeguard at your beach (or the ranger at your favorite park) if you can put up posters to remind people to pick up their trash.

Help students create catchy slogans and make their posters.

Day 2—Detergent Substitutes

Materials

- white vinegar
- old newspapers
- a window

Activity

Say to students:

Remember when we learned about the water cycle? All the water on Earth finds its way back to the oceans, so we should do what we can do to keep it clean on the way.

Detergents get things nice and clean, but they can also add things to the water that are hard to get out. Anytime we use something that cannot pollute the water in place of a detergent, we are helping to keep the water clean on its way back to the ocean.

Try this instead of using a detergent to wash a window:

Put a cup of white vinegar in a pail of water. (Vinegar is made from fruit, grain, or sugar.)

Wash the window with the vinegar and wipe it dry with crumpled newspapers.

Wash one of your classroom windows this way and observe the results.

Have students write down the idea to tell their parents, if you wish.

Day 3—Clean Rivers

Materials

- one or two small trees (if you live near a river or a stream that leads to trees and if you want to take on this kind of project)
- encyclopedias and other illustrated reference materials

Activity

Say to students:

One of the quickest ways for water to get back to the ocean is in a river.

There are two things we can do to keep our rivers clean:

Pick up trash on the river banks.

Plant a tree on the bank of a river or stream. (If this project appeals to you, see if you can get parents to help.)

How do the trees help the river? Have children speculate, based on what they know about the water cycle and erosion.

Day 4—Adopt a Whale

Materials

- encyclopedias and other illustrated reference materials
- a video or book about ocean mammals (*Whales* by Seymour Simon, Thomas Y. Crowell, 1989, is a beautiful book.)

Activity

Give students some information about whales.

Whales are mammals that live in the ocean.

At one time they were hunted almost to extinction, but the International Whaling Commission has banned whaling in hopes of protecting the whales that are left.

Show the video or read the book aloud, taking time to appreciate and discuss the illustrations.

Ask students if they would like to contact one of the organizations that monitors the ocean's whale population. They will make it possible for you to "adopt" a whale and even send you a picture of your whale when it is sighted.

Think of a way to raise the money to adopt a whale. (Recycling something such as paper or aluminum cans would be appropriate.)

Publicize your project within your school to see if other classes would like to help or to participate.

Day 5—Review and Reflect

Materials

- "Review and Reflect" form (See below.)

Activity

Review and discuss concepts introduced during the week. Help students complete the form. Depending on the age group you are teaching, you may want students to dictate their answers to you or to an aide.

Review and Reflect

Name ______________________________ Date ______________

What I Learned:

We need to keep water clean because ______________________________

____

____

______________________________.

The things I have done or plan to do to keep our water clean are ______________

____

____

______________________________.

Our class would like to adopt a whale because ______________________________

____

____

______________________________.

Background

Young children tend to get upset about threats to the environment. Try to keep the focus on things that they can do to help without dwelling on scary pollution statistics and dire predictions.

Day 1—Soil and Erosion

Materials

- same-size tin cans with both ends removed, one for each group
- watering cans or plastic bottles
- water
- scratch paper
- pens or pencils
- watch with a second hand
- outside areas with different types of soil
- Classroom Chart (See below.)

Activity

Remind students that soil can be worn away by running water. Ask who remembers what this process is called. (erosion)

The quicker water is absorbed into the soil, the less erosion takes place. This experiment will show where the chance of erosion is least and greatest.

Divide the class into groups and give each group a can, a supply of water, a watch with a second hand, a piece of scratch paper, and a pen or pencil.

Assign each group a location for conducting their experiment. Depending on the age group and your situation, you may want to send them out to make their observations independently, or you may want to all walk around together, with one group at a time doing the experiment and the rest of the class watching. (That way, you need only one watch!)

Give students these instructions: (You may want to model them first.)

> Press one open end of the can about one inch (2.5 cm) into the soil. (Show them an inch. If using the metric system, make conversions as necessary.)
>
> Quickly fill the can to the brim with water and say "Go!"
>
> The person with the watch counts the seconds it takes the water to soak into the soil.
>
> The person with the scratch paper and pen writes down the number of seconds and the name of the location.

After all the groups have completed the experiment, go back to the classroom and record the data on the chart, using the headings below. Continue on Day 2.

Classroom Chart

Group	Location	Type of Soil	Time to Drain

Day 2—Soil and Erosion *(cont.)*

Materials

- Classroom Chart with recorded data

Activity

Say to students:

We're going to draw some conclusions about what we learned yesterday by looking at the chart we made. To "draw conclusions" means to figure out something by thinking about information that we have.

Ask students to draw conclusions about the following:

Which location should have the least erosion? What makes you think so? Group____, tell us what you observed.

Which location should have the most erosion? What makes you think so? Group____, tell us what you observed.

Look at the chart and check. What was the soil like in the different locations? What can you say about the connection between the condition of the soil and erosion? What could someone do to the soil to prevent erosion?

Discuss.

Day 3—Recycling

Materials

- information about recycling centers in your area
- two large cardboard cartons
- art materials

Activity

Say to students:

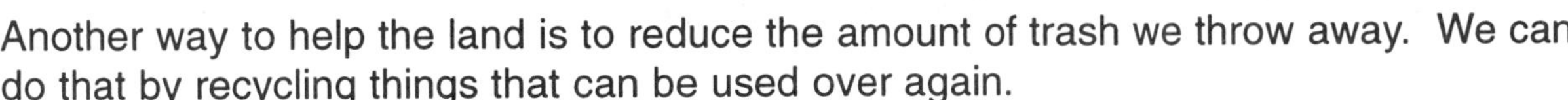

Another way to help the land is to reduce the amount of trash we throw away. We can do that by recycling things that can be used over again.

I have information for all of you to take home about recycling centers in this area.

But before I pass it out, let's talk about what we can do here at school. We can't recycle everything, but we can start with a couple of things and see how it goes. What two materials could we recycle from things we throw away in class and from our lunches? (Try paper and aluminum cans.)

When we have enough materials collected, I will take them to the recycling center.

We can use the money we make to buy something for our classroom.

Have students label and decorate boxes and start to save paper and cans.

Day 4—Junk Art

Materials

items such as the following:

- plastic packing material
- broken toys
- game pieces
- old jewelry
- boxes to store junk

Activity

Say to students:

Things do not have to be collected and sold to be recycled. We can, for example, collect materials and figure out a way to use them for our art projects.

We can clean the closets and shelves and find things right here in our classroom. Then you can look around at home to find other things to add to our collection.

Have students decide on categories of junk and label and arrange boxes.

Day 5—Review and Reflect

Materials

- "Review and Reflect" form (See below.)

Activity

Review and discuss concepts introduced during the week. Help students complete the form. Depending on the age group you are teaching, you may want students to dictate their answers to you or to an aide.

Review and Reflect

Name ______________________________ Date ______________

What I Learned:

We can help the land by learning how to stop ______________________________.

It is least likely to happen when ______________________________.

We can also help by ______________________ instead of throwing items away.

Helping the Animals

Background

Young children tend to get upset about threats to the environment. Try to keep the focus on things that they can actually do to help without dwelling on scary pollution statistics and dire predictions.

Day 1—Our National Parks

Materials

- encyclopedias and other illustrated reference materials
- a selection of brochures about national parks for each group (contact a travel agency)

Activity

Have students meet in groups. Distribute brochures and give students plenty of time to look at and discuss them. *Tell students the following:*

> Our national parks were originally established for people so that the natural wonders of this country would always be there for people to enjoy.
>
> They have become sanctuaries for thousands of wild animals because no one is allowed to hunt and kill wild animals in the national parks.
>
> Has anyone here ever been to one of our national parks? Tell us about it. What animals did you see?
>
> Now that you have looked at the brochures that tell about the parks, write a paragraph telling which one you would most like to visit and why. (Depending on the age group you are teaching, you may want students to dictate their paragraphs to you or to an aide.)

Day 2—Don't Feed the Bears

Materials

- encyclopedias and other illustrated reference materials
- brochures about Yellowstone Park, one for each group (contact a travel agency)

Activity

Have students meet in groups. Distribute brochures and give students time to think about and discuss Yellowstone Park. *Tell students the following:*

> One of the biggest problems at Yellowstone Park is keeping people from feeding the bears. Why do you think people should not feed the bears? Write a paragraph giving your opinions. (Depending on the age group you are teaching, you may want students to dictate their paragraphs to you or to an aide.)

Day 3—The New Zoos

Materials

- encyclopedias and other illustrated reference materials
- a selection of brochures about nearby zoos for each group (contact a travel agency)

Activity

Have students meet in groups. Distribute brochures and give students plenty of time to look at and discuss them.

Say to students:

> At one time, zoos were designed for the comfort of the people who attended them, not for the comfort of the animals that lived there. Animals were kept by themselves in small cages or enclosures rather than given more natural places to live.
>
> The new zoos are designed to keep animals happy. They are kept in places that are as much as possible like their natural surroundings and are together with others of their own kind.

Have each group select a zoo from those described in their brochures and report orally to the rest of the class on the features that impressed them.

Day 4—Don't Feed the Animals

Materials

- encyclopedias and other illustrated reference materials

Activity

Say to students:

In old-style zoos people used to have a great time feeding the animals. They threw peanuts and other food through the bars of their cages.

If you are allowed to feed the animals in the new zoos, you must use the zoo's specially prepared food. Even then, you can feed only a few of the less delicate animals.

Some animals eat only one food. Find out where these animals live and what they must have to eat.

Special Food for Special Animals

Koalas . . . live in ______________________________.

eat only ______________________________.

Giant Pandas . . . live in ______________________________.

eat only ______________________________.

Day 5—Review and Reflect

Materials

- "Review and Reflect" form (See below.)

Activity

Review and discuss concepts introduced during the week. Help students complete the form. Depending on the age group you are teaching, you may want students to dictate their answers to you or to an aide.

Review and Reflect

Name ______________________________ Date ______________

What I Learned:

National parks are as important to animals as they are to people because ______________

____

____.

Zoos keep animals healthy and happy by ______________________________

____

____.

I think it is important to protect wild animals because ______________________

____

____

____

____

____.

Background

Young children tend to get upset about threats to the environment. Try to keep the focus on things that they can actually do to help without dwelling on scary pollution statistics and dire predictions.

Day 1—*The Great Kapok Tree*

Materials

- encyclopedias and other illustrated reference materials
- *The Great Kapok Tree* by Lynne Cherry (Harcourt Brace & Company, 1990)

Activity

Read *The Great Kapok Tree* to your class. It is a book about the community of animals that lives in one giant tree in the Brazilian rain forest. Take time to enjoy the illustrations and discuss the story.

Day 2—The Search for Food and Fuel

Materials

- encyclopedias and other illustrated reference materials
- solar oven (see illustration, page 149)
- small cardboard box
- aluminum foil
- bread and cheese slices to make one or many sandwiches, as desired

Activity

Say to students:

> Often places in the world are not as simple as they first may seem. The rain forest is one such place. It may seem as if taking care of the rain forest plants and animals would be as simple as leaving everything alone. But people need to take care of themselves and their children as well as the animals that share their environment.
>
> In many places, trees in the rain forest are cut down to provide cooking fuel as well as to clear space for growing food crops that people need.
>
> If people had another source of fuel, they would cut down fewer trees. Solar power can be used for cooking under the right circumstances. Solar power is energy directly from the sun.
>
> We are going to make and use a solar/reflector oven and cook something in it.
>
> First, we will line a box with aluminum foil.
>
> Then we will lay a slice of bread on the foil and a slice of cheese on top of the bread.
>
> We will put the box where the sun will shine on it.
>
> We will check every few minutes until the sandwich is done.

Optional: Make enough sandwiches for students to share.

Day 2—The Search for Food and Fuel *(cont.)*

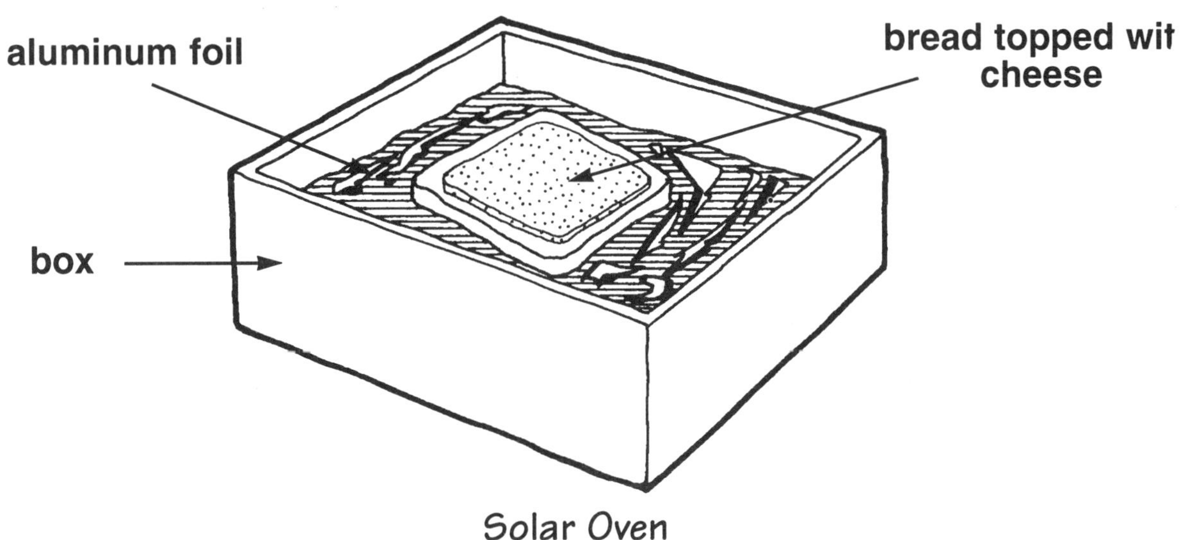

Solar Oven

Day 3—Advantages and Disadvantages of Solar Power

Materials

- outline of chart below sketched on chalkboard

Activity

Say to students:

We have had only one experience with solar power, but let's see what we learned from it. What are the good things about solar power? We will call them "advantages." What are the bad things? We will call them "disadvantages."

Use the information on the chart below for your own convenience and to stimulate discussion. Add information of your own.

Solar Power	
ADVANTAGES	DISADVANTAGES
Once you have the oven, it is free.	It won't work at night.
You don't have to use wood to make it work.	It won't work in the rain.
	It is very slow.

Day 4—Products of the Rain Forest

Materials

- encyclopedias and other illustrated reference materials
- some products of the rain forest, e.g., Brazil nuts

Activity

Say to students:

If the people of the rain forest could make money from rain forest products, they could buy fuel and food without cutting down the trees.

People around the world enjoy eating Brazil nuts and products in which they are used, such as cookies and candy.

Scientists have discovered that many important medicines can be made from rain forest plants.

What would you do if you were a child living in the rain forest? What would you want us to do? (Have an oral discussion about these topics.)

Day 5—Review and Reflect

Materials

- "Review and Reflect" form (See below.)

Activity

Review and discuss concepts introduced during the week. Help students complete the form. Depending on the age group you are teaching, you may want students to dictate their story to you or to an aide.

Review and Reflect

Name ______________________________ Date ______________

On the lines below, briefly tell the story of *The Great Kapok Tree* in your own words.

____

____

____

____

____

Background

Take this opportunity to review and reinforce the concepts and vocabulary introduced during the preceding four weeks.

Day 1—Helping the Oceans

Materials

- none

Activity

Ask students:

Who can tell us one of the things we talked about doing to help the oceans? (pick up your own trash, pick up a little bit extra, put up posters to remind people to pick up their trash, use substitutes for regular detergents, plant a tree on the bank of a river or stream)

We also talked about (or are in the process of) adopting a whale. (Discuss your progress with that project.)

Day 2—Helping the Land

Materials

- Classroom Chart from Week 47, Days 1 and 2
- classroom recycling boxes

Activity

Say to students:

When we talked about helping the land, we concentrated on two things. Can anybody tell me what they were? (erosion and recycling)

Use the chart to describe our erosion experiment. (See Week 47, Day 1.)

What conclusions did we reach? (Answers will vary.)

We decided to do some recycling too. How is that coming along? (Have students check the boxes and report on progress. If you have already taken some things to the recycling center, tell the class how much money they now have. Students may want to discuss what they want to spend the money on when they have enough to make a purchase.)

Day 3—Helping the Animals

Materials

- none

Activity

Ask students:

Why were our national parks originally established? (They were established so that people could always enjoy the natural wonders of the United States.)

How have wild animals benefitted from the parks? (They have become sanctuaries for wild animals.)

Have students share their paragraphs about the dangers of feeding bears in the wild.

Discuss with students:

What did we learn about the new zoos? (The new zoos are designed to keep animals happy. The animals are kept in places that are as much as possible like their natural surroundings and are together with others of their own kind.)

Who remembers what they found out about the koala? Where does it live? (Australia) What does it eat? (eucalyptus leaves)

What about the giant panda? Where does it live? (China) What does it eat? (bamboo)

Are people allowed to feed the animals in the new zoos? (only some animals with food provided at the zoo)

Day 4—Helping the Rain Forests

Materials

- *The Great Kapok Tree* by Lynne Cherry (Harcourt Brace & Company, 1990)

Activity

Say to students:

What book did I read to you about the rain forest? (*The Great Kapok Tree*)

What was it about? (the community of animals that lives in just one giant tree in the Brazilian rain forest)

What did you enjoy most about this book? (Answers will vary.)

Why do people cut down the rain forest trees? (In many places the trees in the rain forest are being cut down to provide cooking fuel as well as to clear space for growing food crops that people need.)

What did we learn about solar power for fuel? (Answers will vary.) What experiment did we do? (solar/reflector oven)

What else did we learn about the rain forest and its people? (If the people of the rain forest could make money from rain forest products, they could buy fuel and food without needing to cut down the trees.)

Day 5—Review and Reflection

Materials

- "Review and Reflect" form (See below.)

Activity

Give students necessary help to complete the form below. Depending on the age group you are teaching, they can do it themselves or dictate their information to you or to an aide.

Review and Reflect

Name ______________________________ Date ________________

Choose words from below to complete this story.

Students can help the Earth in many ways. We can help to keep the ocean clean by picking up ______________________. We can help the land by learning how to keep running water from causing ______________________ and by ______________ things to be used again instead of just throwing them away. We can help animals by being interested in our national ______________________ and by visiting the zoos that try to keep animals happy and healthy in places that are like their natural ______________________.

We can help the rain forests by using things that are ______________________ there.

grown	trash
environments	erosion
parks	recycling

The Reflections Component

Part of the New Assessment

Reflections have become an integral and very valuable part of the new assessments in all curricular areas. We probably associate them primarily with the writing process and with portfolios. They are a perfect match for portfolio assessment because they are designed to look at progress, and portfolios provide a record of progress over a period of time. Reflections were originally associated with the writing process because writing samples made up the bulk of the first student work to be preserved systematically in portfolios.

Possible for Primary Students

Reflections, which were once thought to be the province of older, more sophisticated students are also quite possible for students in the primary grades. Younger students can be introduced to the reflections experience with forms that prompt their responses, or they can be encouraged to reflect orally while the teacher records their thoughts in writing.

An Opportunity to Think About Thinking

Reflections are becoming recognized as an important part of the new assessment process in science because they provide a valuable opportunity for students to think about thinking. Thinking about thinking, or metacognition, was a major educational focus a few years ago. Although not as popular now, the idea is still fresh and important. If students are given information about the process of thinking and an opportunity to apply that information to their own thinking, they get better at it. Since scientific investigations are based on the development of thinking skills, reflecting on scientific investigations brings an automatic bonus—thinking about thinking.

It is interesting to note that even the most simplified thinking models can give students a real advantage in developing their own thinking. Cognitive psychologists believe that simply being aware that their brains continually sort and report incoming information by building new links to connect the data helps students enjoy adding and retrieving ideas from their store of knowledge. Whether or not the model is physiologically accurate, just having one seems to work because it gives students a new and welcome feeling of self-awareness and self-direction.

Make Time for Reflection

Make time for student reflection at the end of each investigation and again before conference times and at the end of the school year. Let students browse through their collected work, note their own improvement and growth, and claim ownership of their progress.

This is an example of how to use a form to help students start the process of reflecting on their own work in science. This particular form was designed for primary students and requires little writing. Review the science skills listed on the form before you start. Allow plenty of time to look over the work that is being reflected upon. When the form is completed, attach it to the work and include it in the student's portfolio.

Reflections on Science

Name Pete Jones Date 6/1/97

When I look back at the work I have done, I feel

EXAMPLE

Check (✔) all that apply.

I have gotten better in
- [x] observing.
- [x] collecting information.
- [ ] using scientific tools.
- [ ] classifying.
- [ ] inferring.
- [x] communicating.

I am really proud of my magnets investigation.

Next time I do an investigation I will use a magnifier.

Reproduce this form for your students to use as they start the process of reflecting on their own work in science. This particular form was designed for primary students and requires little writing. Review the science skills listed on the form before you start. Allow plenty of time to look over the work that is being reflected upon. When the form is completed, attach it to the work and include it in the student's portfolio.

Reflections on Science

Name ______________________________ Date ______________

When I look back at the work I have done, I feel

Check (✔) all that apply.

I have gotten better in
- ☐ observing.
- ☐ collecting information.
- ☐ using scientific tools.
- ☐ classifying.
- ☐ inferring.
- ☐ communicating.

I am really proud of ____

____.

Next time I do an investigation I will ______________________________

____.

Most primary students will do better in reflecting on their own progress with the help of a loosely structured form such as the one on page 156. However, some students may be sophisticated enough to try an essay. You may want them to reflect orally while you write down their comments.

The Reflective Essay

Teacher Script

(Pass out portfolios to the class and have the students follow these instructions.)

Say to Students:

- Look through the work in your science portfolio and put it in order by date.
- Take time to think about the changes you can see. This is called "reflecting."
- Think about things you have done, such as the following:
  - observing
  - collecting information
  - using scientific tools
  - classifying
  - inferring
  - communicating
- What do you know that you did not know at the start of the year? What can you do better? What would you still like to do better?
- Write as much as you can about your own progress.

The Reflective Essay

Student Prompt

Name ______________________________ Date ______________

- Look through the work in your science portfolio and put it in order by date.
- Take time to think about the changes you can see. This is called "reflecting."
- Think about things you have done, such as the following:
  - ✔ observing
  - ✔ collecting information
  - ✔ using scientific tools
  - ✔ classifying
  - ✔ inferring
  - ✔ communicating
- What do you know that you did not know at the start of the year?
  What can you do better? What would you still like to do better?
- Write as much as you can about your own progress.

Index